THE COMMUNICATION
LANGUAGE IMPAIRED C

THE COMMUNICATION OF LANGUAGE IMPAIRED CHILDREN

A study of discourse coherence in conversations of
Specific Language Impaired and Normal Language Acquiring
children with their primary caregivers

Hans van Balkom

Institute for Rehabilitation Research, Hoensbroek,
and the Dutch Organisation for Applied Scientific
Research TNO Delft, The Netherlands

SWETS & ZEITLINGER B.V. AMSTERDAM / LISSE

PUBLISHING SERVICE

The Institute for Rehabilitation Research (IRV) was established in 1981 and has close ties with the University of Limburg, Maastricht; The Lucas Foundation for Rehabilitation, Hoensbroek; and the Dutch Organisation for Applied Scientific Research TNO. The research conducted at the IRV is multidisciplinary in nature. It builds on the interface between the various medical and paramedical disciplines which are involved when addressing problems related to the individual's disability.

The IRV Series in Rehabilitation Research covers topics from the IRV's research programme, including studies on the rehabilitation process, augmentative communication and technology for independent living.

Hans van Balkom received his degree in 'General Linguistics' from the University of Amsterdam, The Netherlands. Dr. van Balkom is primarily involved in conducting research on 'Language Pathology' and 'Augmentative and Alternative Communication (AAC)'. He is a research-coordinator of the 'AAC Research & Development Programme' at the Institute for Rehabilitation Research in Hoensbroek (The Netherlands) as a staff-member of the Dutch Organisation for Applied Scientific Research TNO.

Library of Congress Cataloging-in-Publication Data

(applied for)

Cip-gegevens Koninklijke Bibliotheek, Den Haag

Balkom, Hans van

The communication of language impaired children : a study of discourse coherence in conversations of Specific Language Impaired and Normal Language Acquiring children with their primary caregivers / Hans van Balkom.- Amsterdam [etc.] : Swets & Zeitlinger.- Ill., fig., tab. - (IRV Series in rehabilitation research, ISSN 0925-8396 ; 2)
Ook verschenen als proefschrift Universiteit van Amsterdam, 1991 - Met lit. opg.
- Met samenvatting in het Nederlands.
ISBN 90-265-1183-3
NUGI 725
Trefw.: kindertaal ; onderzoek /spraakstoornissen ; kinderen.

Omslagontwerp: Rob Molthoff
Druk omslag: Casparie, IJsselstein
Druk: Offsetdrukkerij Kanters B.V., Alblasserdam

© 1991 Swets & Zeitlinger B.V., Amsterdam/Lisse

All rights reserved. No part of this publication may be reproduced, stored in a retrieval system, or transmitted, in any form or by any means, electronic, mechanical, photocopying, recording, or otherwise, without the prior written permission of the publisher.

ISSN 0925 8396
ISBN 90 265 1183 3
NUGI 725

To Agnes, Rachel and Judith

Preface

By watching and listening to young children's conversations with their parents (or other caregivers) we learn how children acquire linguistic abilities as a useful tool to influence and control events around them. The rapid and complex process of language acquisition and the development of communicative abilities is guided by the parents. The cradle of all language use is found in the early conversations between the child and his parents. Parents (and other caregivers) adapt their language behaviour when interacting with babies and young children. They also exhibit specific nonverbal acts in the conversations with their children (such as facial expressions, frowning, extended eye-contact, vocalisations). Parents often synchronise their conversational turns with what their child is doing or is likely to do next in such a way that the child's actions are seen as coherent activities and responses. Several studies have indicated the importance of coherent, fluent information exchange in conversations (or discourse) between caregivers and their children. Coherent or connected caregiver-child discourse is mentioned as the vehicle for a successful and smooth development of language and communicative abilities.
In most studies on the communicative interaction between caregivers and their language impaired child the establishment of coherent discourse is seen as the main problem in language and communication development. Besides the already existing language problems of language impaired children, disrupted and nonfluent conversations between the children and their caregivers worsens the premises for language and communication developments. Unfortunately, the number of studies published regarding discourse coherence in conversations between caregivers and their language impaired children is surprisingly limited and controversary in the methods used and the results discussed.

This book presents the results of a longitudinal study which focuses on differences in the use of some specific means for creating discourse coherence in conversations between twelve language impaired children, six normal language acquiring children and their caregivers. All children were at age

2;6-3;0 at the start of the study and at age 4;0-4;6 at the end of the study.
The specific aspects of discourse coherence investigated touch upon '**turn-taking**' (e.g. the number and mean length of turns), '**child adjusted register**' (e.g. the number of child directed and imitated utterances of the caregiver), '**topic management**' (e.g. the number and mean length of conversational themes initiated by the caregiver and child), '**communication acts**' (e.g. the variety of communication acts used), '**communication breakdowns and repairs**' (e.g. the number of unfinished utterances, parallel talk and corrections used). The main research issues concerning the study of discourse coherence, covered by these five areas, are formulated in a set of twenty-four hypotheses. The data for this study were derived from a broader, longitudinal investigation of language and communication development, aimed at constructing a reliable, standardised and clinically useful linguistic and pragmatic method for Interaction Analysis of Communicative Abilities (IACV).

The study reported in this book presents a conceptual framework for studying morphosyntactical and pragmatic aspects in the communicative (verbal and nonverbal) behaviour of the caregivers, their language impaired and their normal language acquiring children. After a general introduction of the study in chapter one, the chapters two and three discuss the ways in which discourse coherence is or can be studied. The chapters four and five consider the definitions of the categories used, the set up of the research method and describe the hypotheses. In chapter six the results are presented. Chapter seven provides the general discussion of the results.

Hans van Balkom, March 1991

Contents

Preface

1	**Introduction**	1
1.1.	General Purpose of the Study	2
1.2.	General Outline of the Study	4
1.3.	The Need for a Specific Research Model	5
1.4.	Specific Language Impairments (SLI)	6
1.5.	Pragmatics	7
1.6.	Interaction, Communication and Language Representation	8
1.7.	Text, Conversation, Discourse	10
1.8.	Coherence and Cohesion	11
1.9.	Organisation of the Book	12
2	**The study of discourse coherence**	15
2.1.	Basic Approaches	16
2.1.1.	Sociology and Ethnomethodology	16
2.1.2.	Sociolinguistics and Textlinguistics	16
2.1.3.	Relevancy of Conversation and Discourse Analysis	17
2.2.	Models for Studying Discourse Coherence	18
2.3.	Generative Models of Discourse Coherence	18
2.3.1.	Global Models of Discourse Coherence	19
2.3.2.	Local Models of Discourse Coherence	21

2.3.2.1.	Lexico-syntactic Models	22
2.3.2.2.	Illocutionary Models of Discourse Coherence	23
2.4.	Descriptive Models of Discourse Coherence	27
2.5.	Alternatives to Generative and Descriptive Models	28
2.5.1.	Situation-Interpreter based Approaches	29
2.5.2.	Cognitively Grounded Conversational Rules Model	30
2.6.	Summary	31
3	**Coherence in adult–child conversations**	**35**
3.1.	Coherence: the shaping of discourse	35
3.2.	Current status of interaction studies with Specific Language Impaired (SLI) children	39
3.2.1.	Turn-taking	40
3.2.2.	Child Adjusted Register (CAR)	44
3.2.3.	Topic and Theme Management	47
3.2.4.	Communication Acts	50
3.2.5.	Communication Breakdown and Repairs	51
3.3.	Summary	53
4	**Interaction analysis of communicative abilities (IACV)**	**55**
4.1.	Adult-child discourse in IACV	56
4.2.	Units of Description and Analysis in IACV	56
4.2.1.	Situation and Context	58
4.2.2.	Verbal and Nonverbal Acts	58
4.2.3.	Exchanges and Moves	62
4.2.4.	Turns, Turn-exchanges and Backchannels	63
4.2.5.	Subsequences and Themes	64
4.2.6.	Topic of discourse	65
4.3.	Coding Discourse Coherence in IACV	67
4.3.1.	Initiations	67
4.3.2.	Responses and Comments	70
4.3.3.	Ellipsis	72

4.3.4.	The Relation with Previous Acts	72
4.3.5.	Imitations	73
4.3.6.	Lexical Coreference	74
4.3.7.	Communicative Functions	75
4.4.	Coding Discourse Incoherence in IACV	77
4.4.1.	Unfinished Verbal Acts	78
4.4.2.	Simultaneously occurring Acts or Overlaps	79
4.4.3.	Incorrect Language Use	80
4.4.4.	Consecutive initiations	81
4.5.	Hypotheses	82
4.5.1.	Turn-taking	82
4.5.2.	Child Adjusted Register (CAR)	83
4.5.3.	Topic and Theme Management	84
4.5.4.	Communication Functions	85
4.5.5.	Communication Breakdowns and Repairs	86
4.6.	Summary	88
5	**Research method**	**91**
5.1.	Introduction	91
5.2.	Selection procedures	92
5.2.1.	Chronological age and family background	93
5.2.2.	Subjects	94
5.2.3.	Language production and language comprehension	98
5.2.4.	Hearing abilities	98
5.2.5.	Psycho-social functioning	98
5.2.6.	Medical history	99
5.3.	Situation and Equipment	99
5.3.1.	Transcription	100
5.3.2.	IACV Analysis Procedure and Coding Reliability	101
5.3.3.	Compiling the data	102
5.4.	Summary	103

6	**Results**	**105**
6.1.	Total Number of Acts used by Adult and Child	106
6.2.	Turn-taking	109
6.2.1.	Backchannelling	109
6.2.2.	Ellipsis	111
6.2.3.	Mean Length of Turn (MLT)	111
6.2.4.	Initiations Adult	115
6.2.5.	Reinitiations Adult	115
6.2.6.	Self-related Imitations Adult	116
6.2.7.	Partner-related Imitations Child	117
6.2.8.	Nonverbal Initiations	120
6.2.9.	Summary	122
6.3.	Child Adjusted Register (CAR)	123
6.3.1.	Mean Length of Utterance (MLU)	123
6.3.2.	Self-related Imitations Adult	124
6.3.3.	Requests for Information Adult	125
6.3.4.	Summary	125
6.4.	Topic and Theme Management	126
6.4.1.	Introduction New Theme	126
6.4.2.	Mean Length of Subsequence (MLS)	128
6.4.3.	Partner-related Imitations Child Between Groups Effect	129
6.4.4.	Self-related Imitations Adult	129
6.4.5.	Summary	130
6.5.	Communication Functions	130
6.5.1.	Requests for Information Adult and Child	130
6.5.2.	Communication Functions Adult and Child	131
6.5.3.	Summary	134
6.6.	Communication Breakdowns and Repairs	135
6.6.1.	Unfinished Verbal Acts	135
6.6.2.	Parallel Talk	136
6.6.3.	Adult Clarification Requests	136
6.6.4.	Adult Corrections of Previous Act	137
6.6.5.	Consecutive Initiations	138
6.6.6.	Faulty Responses	138
6.6.7.	Faulty Initiations	138
6.6.8.	Summary	140

7	Discussion	141
7.1.	Total Number of Acts used by Adult and Child	141
7.2.	Turn-taking	143
7.2.1.	Backchannelling	143
7.2.2.	Ellipsis	144
7.2.3.	Mean Length of Turn (MLT)	145
7.2.4.	Initiations Adult	146
7.2.5.	Reinitiations Adult	147
7.2.6.	Self-related Imitations Adult	148
7.2.7.	Partner-related Imitations Child	148
7.2.8.	Nonverbal Initiations	149
7.2.9.	Summary	149
7.3.	Child Adjusted Register (CAR)	150
7.3.1.	Mean Length of Utterance (MLU)	150
7.3.2.	Self-related Imitations Adult	151
7.3.3.	Requests for Information Adult	151
7.3.4.	Summary	152
7.4.	Topic and Theme Management	152
7.4.1.	Introduction New Theme	152
7.4.2.	Mean Length of Subsequence (MLS)	153
7.4.3.	Partner-related Imitations Child	154
7.4.4.	Self-related Imitations Adult	154
7.4.5.	Summary	155
7.5.	Communication Functions	155
7.5.1.	Requests for Information Adult and Child	155
7.5.2.	Communication Functions Adult and Child	156
7.5.3.	Summary	157
7.6.	Communication Breakdowns and Repairs	157
7.6.1	Unfinished Verbal Acts	157
7.6.2.	Parallel Talk	157
7.6.3.	Adult Clarification Requests	158
7.6.4.	Adult Corrections of Previous Act	158
7.6.5.	Consecutive Initiations	159
7.6.6.	Faulty Responses	159
7.6.7.	Faulty Initiations	160
7.6.8.	Summary	161

7.7.	Overall Summary of the Results	161
7.7.1.	The Participation in Discourse Coherence	162
7.7.2.	The Role of the Caregiver	164
7.7.3.	Grammatical Development and Pragmatic Skills	168
7.7.4.	Implications for Further Research	169
8	**BIBLIOGRAPHY**	**171**

APPENDICES

A	Transcript Conventions in IACV	189
B	IACV Transcript Example	193
C	IACV Communication Functions (definitions and examples)	207
D	Interjudge Reliability Measures for the IACV Pragmatic Categories (based on Cohen's 'Kappa', 1960)	235
E	Mean and Standard Deviations of Communication Function Categories of the Caregivers	239
F	IACV Tables of Statistics (Tables F1a,b to F6 belonging to the chapters 6 and 7)	245
G	Early Social Communicative Scales (ESCS) (adapted version in Dutch)	261

CHAPTER 1

Introduction

This chapter discusses the main purpose and general outline of the study, investigating presupposed differences in discourse coherence of conversations between caregivers* and their Normal Language Acquiring (NLA; N=6) and Specific Language Impaired children (SLI; N=12) at preschool age (from 2;6** to 4;0). The study is primarily concerned with how NLA and SLI children initiate and sustain conversations with their caretakers. Discourse coherence is studied from the perspective of both conversation partners (see chapter 4). In order to avoid an undifferentiated use of key terms and notions, such as language, interaction and communication, conversation and discourse, coherence, coherency and cohesion, Specific Language Impairments and pragmatics, definitions are given in this chapter. More specific terms and definitions, used for the coding of discourse in coherence in this study, are given in chapter 4.

---

* Throughout this book 'adult', 'caregiver' and 'caretaker' are used as synonyms, referring to the adult communication partner of the child.
** (years;months)

1.1. General Purpose of the Study

The analysis of language used in communicative interaction between SLI children and their caregivers has implications for both basic research and clinical methods. Specific Language Impaired (SLI) children show significant deficits in their linguistic functioning and have appropriate nonverbal intelligence, normal hearing, and a seemingly normal medical or developmental history (section 1.4.). Most of the recently published studies on language development of exceptional populations are based on the assumption that adult-child exchanges are the result of mutual cooperation between caregiver and child (Dromi, 1989; McTear & Conti-Ramsden, 1989, 1990).

The presence of an exceptional condition (e.g. SLI, hearing impairment, blindness, mental retardation, cerebral palsy) in the dyad required adaptations related to the specific characteristics of the producer of a message as well as of its recipient (Conti-Ramsden & Friel-Patti, 1983; Dromi & Beny-Noked, 1984). These studies indicated that the communication partners used specific adjustments for establishing and maintaining successful transactions. Most of these modifications are made in order to achieve a connected, coherent information flow between adult and child (Day, 1986; Chaika & Lambe, 1989; Junefelt, 1989; Evans, 1989; Heim, 1989). From a clinical point of view, it is important to examine the linguistic environment of SLI children since it may include aspects which do not facilitate language development. If the communicative interaction of caregivers with their SLI children shows deviant patterns, these patterns may contribute to the perseveration of the language disability (Schodorf & Edwards, 1983). Descriptive evidence on the contributing factors to the problem can therefore help in identifying specific goals for intervention, which will sometimes incorporate the modification of parental verbal and nonverbal interactions with their children (Rieke, Lynch & Soltman, 1977; Schachter, 1979; Lasky & Klopp, 1982; Russo & Owens, 1982; Dromi, 1984; McTear & Conti-Ramsden, 1989).

The number of published studies on adult-SLI child interaction has been surprisingly limited and controversary in the methods and results discussed (Fey & Leonard, 1983; McTear, 1985; McTear & Conti-Ramsden, 1989; see also chapter 3). In most of the reported studies which focus on the communicative interaction between the SLI child and his caregiver, the establishment of discourse coherence is mentioned as the main problem (McTear, 1985; Kamhi, 1989; Conti-Ramsden, 1988, 1989). This study will gather more evidence for the discussion of problems in discourse coherence. The study hopes to contribute to a better understanding of the problems SLI children and their caretakers have with discourse coherence.

In summary:
The study focuses on differences in the use of some specific means for creating discourse coherence in conversations between SLI children, NLA children and their caregivers, as is indicated in several studies. The specific aspects of discourse coherence investigated touch upon 'turn-taking' (e.g. the number and mean length of turns), 'child adjusted register' (e.g. the number of child directed and imitated utterances of the caregiver), 'topic management' (e.g. the number and mean length of conversational themes initiated by the caregiver and child), 'communication acts' (e.g. the variety of communication acts used), 'communication breakdowns and repairs' (e.g. the number of unfinished utterances, parallel talk and corrections used).

The literature studied (and reviewed in the next chapters) indicates that most problems in discourse coherence of SLI children and their caregivers can be covered by the five areas just mentioned (Schodorf & Edwards, 1983; Fey & Leonard, 1983; Conti-Ramsden, 1989; Grunwell & James, 1989). Detailed information about these aspects is discussed in the chapter 3.

In order to investigate the various aspects of discourse coherence, the study is based on two research principles. First, the study is based on naturally occurring language of child and adult. These language data (verbal and nonverbal) were the source for the description and the analysis of discourse coherence. Second, conversation or spoken discourse is a joint production which is created on an act-by-act basis. Discourse coherence is managed through the acts of the conversation partners. Coherence is not solely situated in the discourse itself but in the context of the activities and the communicative intentions of the interactants.

Both research principles result in a specific description and method for interaction analysis used for this study of discourse coherence (see section 1.3. and chapter 4).

The analysis of discourse coherence is based on various levels of language description, such as verbal and nonverbal actions, turn-taking and the organisation of discourse in conversational themes (see chapter 4). Information exchange and learning strategies are transmitted in all interactions between child and adult. Coherence of the information given is a prerequisite for successful and fluent communication. The study of discourse coherence and incoherence helps to identify formal and functional cues in communicative interaction that are used for interpretation and information retrieval.

Many studies have indicated that children at preschool age, from 2;6 to 4;0, are in their most sensitive period for acquiring and practising communication skills (Bloom & Lahey, 1978; Bloom, Roscissano & Hood, 1976; Schachter, 1979; Wells, 1981, 1985; McTear, 1985). This phase is characterised by growing social awareness. The child becomes more mobile and involved in different situations and activities. This enables him to meet a variety of persons in different situations. Throughout their preschool years children become more adept at sharing conversational responsibilities and adapting their utterances to the needs of the listener and the situation (Shatz, 1982; Johnston, 1986).

The specific problems SLI children have with creating and interpreting coherent discourse in conversations with their caregivers has been given limited research interest, as will be discussed in chapter 3.

1.2. General Outline of the Study

The framework for this study is derived from a broader investigation, aiming at the construction of a reliable, standardised and clinically useful linguistic method for Interaction Analysis of Communicative Abilities (IACV*):

- twelve SLI and six NLA children at preschool age and one of their parents were selected for participation
- the parents were asked to participate in a longitudinal study for a period of eighteen months. Nine free play sessions were held at two months intervals
- the free play sessions lasted thirty minutes and were video-taped
- a random selection of five minute samples from each video-recorded session was taken for transcription and analysis
- the first five video-taped minutes of each session were left out for transcription
- the transcripts were analysed by using separate procedures for morphosyntactic description of verbal acts of child and adult
- the transcripts were also analysed by using a programme for describing pragmatic (functional) characteristics of verbal and nonverbal acts of adult and child (e.g. turn-exchanges, initiations, communication acts, imitative relations)
- specific hypotheses concerning the differences in discourse coherence between SLI and NLA children were derived from related studies
- the variables used for coding coherence were derived from the available categories in the analysis programme for pragmatics
- all the variables used were obtained from related studies
- the data collected were statistically compiled by using SPSS (Statistical Package of the Social Sciences).

* The Dutch acronym 'IACV' stands for *Interactie Analyse van Communicatieve Vaardigheden'*.

1.3. The Need for a Specific Research Model

The aspects of discourse coherence described in this study indicate presupposed coherence strategies used in skilled communicative interaction. We do not know enough about the nature of conversation and discourse coherence to enable the presentation of a model of the mature adult conversationalist which can be used to measure conversational (and discourse coherence) development in children. Besides this, most studies of discourse coherence have severe limitations, mainly because of methodological restrictions, such as the use of obtrusive observational devices, varying criteria for the selection of subjects, a limited representativeness of samples and size of samples, a preoccupation with verbal behaviour, a neglect of listener/partner role in conversation (McReynolds & Kearns, 1983; Van Balkom & Heim, 1990). In order to deal with most of these limitations and problems the IACV study in general and this discourse coherence study in particular, a specific framework for the analysis of adult-child conversations was designed. Most of the categories included in the IACV discourse framework have already been established and used with normal, not language disordered, speakers. The IACV study was set up in order to find evidence for the use of the discourse framework as a suitable classification for conversations between SLI children and their caregivers. More specifically the discourse frameworks introduced by Wells (1975, 1985), Schachter (1979) and McTear (1985) were taken as a starting point for the set up of the IACV discourse framework. The study is also based upon research issues in Discourse Analysis, Conversation Analysis and Discourse Coherence, concerning turn-exchanges, speech or communication acts and topic management. The way in which these studies are integrated in the IACV discourse framework is discussed in the chapters 2 and 4. The IACV research method is described in chapter 5.

The IACV discourse framework is based on the following suppositions:
- fully specified transcripts (incorporating verbal and nonverbal acts, non-linguistic situation) need to be available
- these transcripts form the basis for the analysis which needs to indicate reciprocal relations of language behaviour of both partners
- the analysis also must distinguish discourse characteristics at macro (conversation theme) and micro (verbal and nonverbal act) level
- the study needs to offer data for the study of developmental aspects in discourse coherence.

These requirements indicate that discourse coherence is considered as a process in which both partners are actively involved and context and nonlinguistic situation are important influencing factors. The fourth mentioned supposition can only be implemented in a longitudinal study.

So far, we have been using several terms rather loosely, such as discourse and conversation, coherence, interaction and communication. The next sections propose definitions of the most important general terms, to be used in this research report. These definitions cannot be given without the use of terms not yet defined, such as utterance, nonverbal act, turn-exchange. These more specific terms will be discussed in chapter 4, because they are strongly related to the way in which the adult-child discourse is studied within IACV.

1.4. Specific Language Impairments (SLI)

There still is no consensus on defining a unified terminology to describe language learning problems in children. At the "First International Symposium on Specific Speech and Language Disorders in Children*", terminology was an important issue.

Among the prerequisites for normal language acquisition are adequate hearing, adequate integration of perceived stimuli, information processing, intelligence, intact sensory and motor control of gestures and manual signs, knowledge about language rules, conventional knowledge about appropriate social behaviour and language use. There is a group of children with serious difficulties in learning to understand and produce language. These difficulties cannot be explained by hearing impairment, overt neurological disorders, mental or intellectual retardation or social deprivation. Historically, various terms have been used to refer to this group of children, including 'delayed language' (Lee, 1966), 'congenital aphasia' (Eisenson, 1972), 'deviant language' (Leonard, 1972), 'language disorder' (Rees, 1973; Bloom & Lahey, 1978), 'specific language deficit' (Stark & Tallal, 1981), 'developmental dysphasia' (Clahsen, 1988), and 'developmental language disorders' (Conti-Ramsden, 1989). The term "language disorder" has been useful as a clinical description for all children with problems in language learning. This concept however led to difficulties in the refinement of defining problems with language development when associated disorders have to be excluded. This created a dilemma in formulating sufficiently precise research questions of aetiology, remediation strategies and prognoses.

Ingram (1961) was among the first in trying to define difficulties in language learning which cannot be explained by mental retardation, hearing impairment, environmental factors, structural and neuromotor causes, or autistic behaviour. He uses the term "Specific Developmental Language Disor-

---

* University of Reading, Reading (UK), March 29 to April 3 in 1987, organised by AFASIC (Association For All Speech Impaired Children).

ders", a definition by exclusion, which makes it likely that we are not dealing with a unique, single and homogeneous condition. The subsequent literature refers to these specific language development problems as disorders with an unknown cause. According to Robinson (1987), there still is a good deal of information on possible or plausible causes, among which are:
- epidemiological facts (e.g. a significantly larger number of boys affected, specific learning difficulties)
- medical causes (e.g. chromosomatical abnormalities, non-accidental injuries, perinatal abnormalities) and
- certain associated medical features (e.g. clumsiness and pathological non-right handedness).

In this study the SLI children are selected according to the definition of specific developmental language disorders, renamed as *"Specific Language Impairments (SLI)"* (according to more recent references, Fey & Leonard, 1983; Kamhi, 1989). The selection procedure for SLI children and their caregivers is described in chapter 5 (section 5.2.).

1.5. Pragmatics

Pragmatics has been described as "the study of the rules governing the use of language in a social context" (Bates, 1976). Yet, simple as this definition may appear, it covers a wide range of linguistic and non-linguistic phenomena which have been traditionally assigned to this area.

Levinson (1983) presents an overview of several definitions of pragmatics:
- the study of relations between language and context which are encoded in the structure of a language
- the study of aspects of meaning which are not captured in a semantic theory
- the study of relations between language and context which support a theory of language understanding
- the study of the appropriate use of language.

None of these definitions is sufficient on its own.

A good characterisation of the term pragmatics is given by McTear:
> "(...) pragmatics is not an additional level of language, to be added on to syntax, semantics and phonology. Rather, it pervades language use at all levels" (1985,7).

According to McTear and Conti-Ramsden (1989) pragmatics should consider the following aspects of language use:

- discourse and conversational skills (e.g. initiating and responding behaviour, turn-taking, topic management)
- the relationships between pragmatic and other levels of language (e.g. morpho-syntax, semantics, phonology)
- situational determinants (e.g. home or clinical situation, furniture, the presence of strangers).

In fact all these definitions can be seen as specifications or operationalisations of the general and simple definition offered by Bates (1976), making it possible to study language as it is functionally used by communication partners. McTear's definition of pragmatics indicates that the study of language use needs to consider relationships between several language levels, morpho-syntax, semantics and phonology. This is especially true in the study of discourse coherence. Discourse coherence is created through the use of several linguistic strategies, acting across language levels, such as the use of ellipsis (a grammatically incomplete utterance, but pragmatically correct and efficient) and imitations or repetitions. Imitations or repetitions mostly are grammatically complete and correct, but at the same time they can be pragmatically incorrect or functionally inadequate due to the delay in information exchange. The study of discourse coherence has much to do with the investigation of conversational adequacy (e.g. the correctness of responses, initiations and the relevancy of the information given).

Conversational adequacy is realised through the exchange of clearly related and expressed information at all language levels (morpho-syntax, semantics and phonology).

Inadequate information or formulations (e.g. the use of unspecified referents, mispronunciations, faulty initiations and responses) often lead to communication breakdowns and consequently to discourse incoherence. In the study of discourse coherence and incoherence all language levels need equal consideration. McTear's (1985) definition of 'pragmatics' justifies this approach.

1.6. Interaction, Communication and Language Representation

Successful communication is a collaborative enterprise. Each participant takes turns as speaker and listener respectively; as such they construct and encode messages so that they both adequately convey the intended meaning and focus on the informational needs and expectations of the other communication partner at a specific point in the information exchange. Simultaneously the partner tries to interpret the code of the message, drawing upon a variety of

sources of information from the environment and past experience. Communication partners engage in a number of interrelated processes:
- translating a personal experience into the semantic structure of their common language
- encoding and decoding of meaning intentions
- integrating verbal, nonverbal and contextual cues.

The major part of the definitions of 'communication', or the way in which messages are encoded in the process of interpersonal exchange of information, indicates a preoccupation of most researchers with verbal communication. The definitions they use often are formulated from the perspective of communication through speech. The speaker or writer is put at the centre of the process of communication.

However, the study of communication needs a model in which the role of each participant is equally considered and in which verbal as well as nonverbal means of communication have to be regarded. From that point of view communication is seen as an active process in which information between two or more parties is exchanged.

The exchange of information is realised through representations of ideas, intentions and meanings in several codes. There are four groups of codes, based on the specific way of encoding and perceiving of the messages (Van Balkom & Welle Donker-Gimbrère, 1988):
- motor-visual codes (e.g. gestures, sign language, mimicry)
- graphic-visual codes (e.g. pictures, photographs, symbol-sets)
- three-dimensional/tactile codes (e.g. shapes, objects, braille)
- acoustic-auditory codes (e.g. natural/synthetic speech, morse).

The kind of codes to be used are based on the preferences, the conventions and situational context of the users. These codes function as a language:

> "A language is a code whereby ideas about the world are represented through a conventional system of arbitrary signals for communications" (Bloom & Lahey, 1978,4).

According to this definition a 'language' can be represented in several codes, to be selected and used on the basis of conventions and functional agreement of the actual users of the codes.

The way in which language is represented in codes differs at the level of linguistic description. From the viewpoint of linguistics only a limited set of codes functions as an actual 'language'. In linguistics a code represents a language if all the signs and sign-combinations belonging to that code are described and accounted for by a grammar. Besides the codes that represent a language, a set of codes exists without a fully specified grammar (e.g. gestures, mimicry, pictures, photographs, traffic signs). Defining language as a representation of codes makes it possible to study language in its main function of communication with equal consideration of verbal and nonverbal acts and differing levels of linguistic description.

The process of information exchange, or communication, evolves during interaction of the conversational partners. The term 'interaction' denotes the mutual effect of actions performed by the partners; actions which have the intention to gather or maintain the other person's attention to shared activities. 'Communication' has often been distinguished form 'interaction' on the grounds that, whereas communication involves an exchange of meanings, interactions can occur without the participants sharing the same meanings (Bullowa, 1973,3). When we use the term 'communication', interaction (verbal and nonverbal) is implied. If information and social exchanges are to be comprehensible and sufficiently ordered (so as to accomplish personal and group goals), they must be coherent. Communicators accomplish discourse coherence through an array of behavioural and interpretative strategies (see the chapters 2 and 3).

So, the definitions of 'communication' and 'discourse coherence' seem to have a direct relation.

1.7. Text, Conversation, Discourse

There are various terms and definitions for indicating stretches of connected information. Among the most frequently used are 'text', 'conversation' and 'discourse'.

Sidner (1980,122) defines discourse as any connected piece of text or spoken language of more than one sentence.

Van Dijk (1977,3) describes discourse from a more restricted perspective, as those utterances which can be assigned textual structure and are acceptable, well-formed and interpretable. Van Dijk's definition is concerned with rules underlying well-formed discourse. Well-formed discourse can be seen as discourse conforming to prescribed rules. This normative and prescriptive use of the terms 'discourse' and 'text' is difficult to apply in a study of naturally occurring language, referred to as 'discourse' or 'conversation'. In the present study conversation or discourse is seen as the use of codes to negotiate meaning. Discourse is the result of the interpretative activities of the conversational partners and their will to produce verbal and nonverbal acts (e.g. pointing, gaze and (head) nods) coherently used in the sequence of the conversational acts. The word 'text' is used in linguistics to refer to any passage, spoken or written, of whatever length, that forms a unified whole. According to Halliday and Hasan (1976) a text is a unit of language use. A text is not a grammatical unit, such as a clause or sentence, and it is not defined by size.

A text is best regarded as a semantic, meaningful unit. A text does not consist of sentences but is realised by sentences. So, in all views of 'text' there is a notion of coherence. A text will be experienced as coherent if the speaker(s) or actor(s) and the hearer(s), co-actor(s) or receiver(s) cooperate in such a way that each verbal or nonverbal act can be interpreted as being connected with the foregoing. We will use the term 'discourse' or 'conversation' to refer to any sequence or combination of verbal and nonverbal acts between a caregiver and his or her child. The term 'text' is used to indicate the language sample (the verbal and nonverbal acts) itself.

The definition of language has shown (section 1.6.) that coherence is seen as an integral part of language use, and that also the definitions language, interaction, communication, conversation, discourse and text are related to coherence.

1.8. Coherence and Cohesion

In the definition of language presented above, interaction, communication, text and discourse coherence were common issues. The various definitions and operationalisations of the term 'coherence', used in various approaches and models (chapter 2) reveal that coherence is seen as the central notion of discourse. The way in which researchers think that discourse coherence is established directly influences the methodology of the studies conducted. According to the literature dealing with discourse coherence (reviewed in chapter 2), we will use the terms '(discourse) coherence' (Craig & Tracy, 1983), 'coherency' (Ochs Keenan & Klein, 1975; Reichman, 1978), 'sustained discourse' (Wells, 1981), 'connectedness' (Hartveldt, 1987) as synonyms. However, the term 'discourse coherence' is preferred. In the current study coherence is seen as a property of the participants' activities which they realise while interpreting and producing discourse.

According to Hartveldt (1987), coherence is not purely a property of a text itself, but of whatever the participants appear to do in constructing and interpreting a text. Conversational partners presuppose coherence in any text. This can lead to misunderstandings between conversational partners, specifically in those situations where large time lapses occur between successive acts.

The larger the time lapse between a verbal or nonverbal act produced and the one it should be connected with, the more clearly this must be signalled by the speaker or actor.

"Make your conversational contribution such as is required, at the stage at which it occurs, by the accepted purpose or direction of the talk exchange in which you are engaged". (Grice, 1975,45)

Halliday and Hasan (1976) use the terms 'texture' and 'cohesion'. A text has texture, and this is what distinguishes it from something that is not a text. It derives this texture from the fact that it functions as a unity with respect to its environment.

There are certain linguistic features which can be identified as contributing to giving a text texture (e.g. ellipsis, imitation, rephrasing, comments, lexical co-reference).

A number of linguists (e.g. Brown & Yule, 1983) have drawn a distinction between cohesion and coherence, and have shown that cohesion does not necessarily lead to the overall coherence of a text (see also section 2.3.2.1., example 1). In the present study we will use the terms 'cohesion' and 'coherence' in order to describe the connectedness of conversation at micro level (within a verbal or nonverbal act) and at macro level (between verbal and nonverbal acts, turn exchanges) respectively. Cohesion is seen as a semantic concept; it refers to relations of meaning that exist within a text, and that define it as a text. Cohesion occurs where the interpretation of some element in the discourse depends on that of another. Cohesion refers to all lexico-grammatical structures used for establishing coherent discourse. We use the term 'cohesion' to refer to lexical and grammatical connections within verbal acts.

The term 'coherence' refers to the connected information in discourse, established between verbal and nonverbal acts of the communication partners and often cohesion may be subsumed.

1.9. Organisation of the Book

So far we have briefly considered general aspects of our study (aims, motives, scope, research model, methods involved and general terminology). The remainder of the book is organised as follows:

Chapter 2 reviews the literature about discourse coherence. As indicated earlier (section 1.1), the development of discourse coherence is of main importance for acquiring conversational skills. Here the relevance of the many approaches and models for studying discourse coherence is discussed in the light of the purpose of our study.

Chapter 3 offers an overview and a discussion of the literature concerning the discourse coherence in adult-child conversations. The literature refers to normal and deviant language acquisition and linguistics in clinical practice.

Chapter 4 presents the set-up of the IACV research method used for analysing discourse coherence. According to the research method used an account is given of the way in which adult and child structure their discourse and how this is studied within IACV. The research hypotheses are described with reference to the literature reviewed in the chapter 3.

Chapter 5 is concerned with the research method. The information given deals with the research design, the selection criteria for the subjects (NLA and SLI children and their caregivers), the observation techniques (data collection), transcription and analyses and the statistical compilation of the data.

Chapter 6 describes the results of the data analysed, referring to the research hypotheses discussed in chapter 4.

Chapter 7 discusses the implications of the results.

Appendices A and B specify and illustrate the IACV Transcript Conventions.

Appendix C summarises the definitions of Communication Functions used in the IACV Pragmatic Analysis.

Appendix D presents the results of the inter-judge reliability study of the codings in the IACV Pragmatics Analysis based on Cohen's 'Kappa' (1960).

Appendices E and F present the tables with statistics (belonging to the chapters 6 and 7).

Appendix G presents the 'Early Social and Communicative Scales (ESCS)' published by Seibert and Hogan (1982).

CHAPTER 2

The study of discourse coherence

As indicated in the previous chapter, the present study is primarily based on theories and models derived from adult-child discourse, studies on developmental pragmatics, studies on normal and deviant language development and language input studies.

Besides these research areas there is a vast number of studies in sociology, sociolinguistics and textlinguistics concerning discourse coherence models and allied theories. The majority of the discourse coherence models is designed for studying cohesion and coherence in narratives. Some models are specifically designed for applications in Artificial Intelligence (e.g. man-machine dialogue systems, automatic translation of texts). Most of the developed discourse coherence models are very difficult to apply to the study of discourse coherence in natural adult-child conversations. This does not imply however that the definitions, concepts and methods which are available through these models have not influenced the studies of discourse coherence in natural conversations.

In this chapter a short review is given of models which have had an effect on the definitions of discourse coherence and the methods used for studying discourse coherence in natural conversations. The way in which discourse coherence in adult-child conversations is investigated cannot be accounted for without reference to the research concerning the construction of discourse coherence models. The final section of this chapter summarises the relevant connections between the reviewed models and the study at issue.

2.1. Basic Approaches

In this section a brief discussion is given of different approaches in the study of language use in conversations (for a more detailed overview see Brown and Yule (1983) and Haft-van Rees (1989)), specifically dealing with theories and models concerning coherence. Roughly speaking, there are two different approaches in studying the use of language in conversations. They mainly differ in research methodology and the disciplines involved.

2.1.1. Sociology and Ethnomethodology
The branch in sociology which is most interested in the study of social interaction between members of a social community is 'ethnomethodology'. The research of ethnomethodologists concerning the study of communicative interaction is referred to as 'Conversation Analysis'.

The study of conversations reveals principles concerning the use of strategies for participating in social interaction and the way in which conversational rules are used and defined intersubjectively.

The most important aim is the description of interactional procedures used by conversational participants while performing and while interpreting their contributions. Reseachers describe the conversational flow step by step according to the organisation imposed by the participants. The theoretical implications from their studies are based on the empirical conversational data observed (referred to as 'observational naivity'). The research is based on basic data (audio and video recordings) and accurate and fully specified transcripts. The main objective of the conversational studies of ethnomethodologists is to reveal information about the specific procedures used to organise and define communicative interactions between members of a specific social community.

Important representatives are Garfinkel (1967); Sachs, Schegloff and Jefferson (1974).

2.1.2. Sociolinguistics and Textlinguistics
Linguistic research focuses mainly on the study of relations between form and functions of language utterances, sequences and combinations of utterances in conversations. Linguistic research concerning the use of language in conversation and written texts is often called 'Discourse Analysis'. The most important

goal of Discourse Analysis is the description of structured principles of conversations and written texts. Researchers try to identify rules and conventions used in order to clarify observed stretches, sequences of utterances and topical structures in spoken and written discourse. The discourse is described as a whole, perceived from the perspective of an outsider (external observer). Speech Act theory and the theory of Textlinguistics (covering grammatical and meaning relations in written discourse) are the leading theories as far as analysing the data is concerned. In written discourse the sources mostly consist of prestructured literal texts. 'Prestructured' means that these texts were all selected or specifically constructed beforehand and for the purpose of the study by the researcher or the authors. In spoken discourse the transcripts are goal directed; irrelevant aspects are left out. In some cases utterances and interactional structures are invented by the researcher. The researcher's assumption of relations between form and function of utterances and rules for the sequencing of utterances are tested. Important representatives of Discourse Analysis are Labov (1970); Searle (1969, 1975); Sinclair and Coulthard (1975); Van Dijk (1985); Edmondson (1981).

2.1.3. Relevancy of Conversation and Discourse Analysis

Conversation Analysis and Discourse Analysis differ with respect to their fundamental aims and the methodology used. Both Discourse and Conversation Analyses give results that enhance the knowledge about conversational structure and stimulate research on conversational language use.

Discourse Analysis as well as Conversation Analysis offer possibilities for a theoretical framework to be used in the study of discourse coherence in child-adult conversations. We will not follow the debate between researchers in Conversation or Discourse Analysis and neither will we choose either of the analyses as explicit framework for our study. The research method used and some aspects of the discourse framework set up in IACV are known from Discourse Analysis (e.g. communicative acts, cohesion and coherence) as well as from Conversation Analysis (e.g. observation and transcription, definitions of turn-exchanges, topic and theme). We use accurate, fully specified transcripts of informal dialogues and try to analyse the transcripts step by step as sequenced by the partners (according to the principles of Conversation Analysis). Discourse coherence is an important object of study in Discourse and Conversation Analysis. The methods used for studying discourse coherence in both Discourse and Conversation Analysis differ in the way they deal with the many regularities and irregularities in spoken discourse. The discourse coherence models, which are reviewed in the next sections, indicate to what extent they handle (ir)regularities in spoken discourse. These models are developed either from the perspective of Discourse Analysis or from the perspective of Conversation Analysis.

2.2. Models for Studying Discourse Coherence

Goldberg (1983) discusses two main approaches in studying discourse coherence based on the assumption that discourse coherence either can be seen as the result of a set of regulative and constitutive rules or as the direct result of the acts and interpretations of the conversational partners. The first approach uses methods and models for studying discourse coherence based on a set of concatenation rules and predictive relations and therefore is referred to as 'Generative Models of Discourse Coherence' (section 2.3).

The term 'generative' indicates the importance of rules and conventions in describing and explaining subsequent verbal and nonverbal acts. The second approach uses Descriptive Models for studying discourse coherence and is referred to as 'Descriptive Models of Discourse Coherence' (section 2.4). The majority of Generative and Descriptive Models are developed from a speaker's point of view. However, it should be stressed that creating discourse coherence is not solely the result of the work done by the speaker, but also by the hearer, who obviously becomes a speaker or actor at his time. The incorporation of speaker, hearer and situation in a model of Discourse Coherence is evident in the so-called 'Situation-Interpreter based Approaches' and the 'Cognitively Grounded Rules' (sections 2.5.1 and 2.5.2). Figure 2.1. gives an overview of the models and approaches to be discussed in the following sections.

2.3. Generative Models of Discourse Coherence

Generative Models are to be divided into two groups, the first of which predicts text samples based on highly structured conversational and textual structures (Global Models) whereas the second group predicts verbal and nonverbal acts which can follow a preceding verbal or nonverbal act (Local Models). Generative Models of Discourse Coherence are based on a number of distinct theories, most notably the Sequencing Rules Models (Searle, 1969, 1975; Sacks, Schegloff & Jefferson, 1974), the Cohesion Model (Halliday & Hasan, 1978), Artificial Intelligence and the Cognitive Rules Model (Hobbs, 1979, 1983; Sinclair, 1985; Polanyi, 1985). Generative Models explain discourse coherence by appealing primarily to discourse features and rules. Generative

Models provide a set of concatenation rules for the production of locution sequences (e.g. adjacency pairs, ellipsis, preconditions), so as to ensure that the propositions they represent will be orderly and meaningfully arrayed.

FIGURE 2.1.   Discourse Coherence Approaches and Models

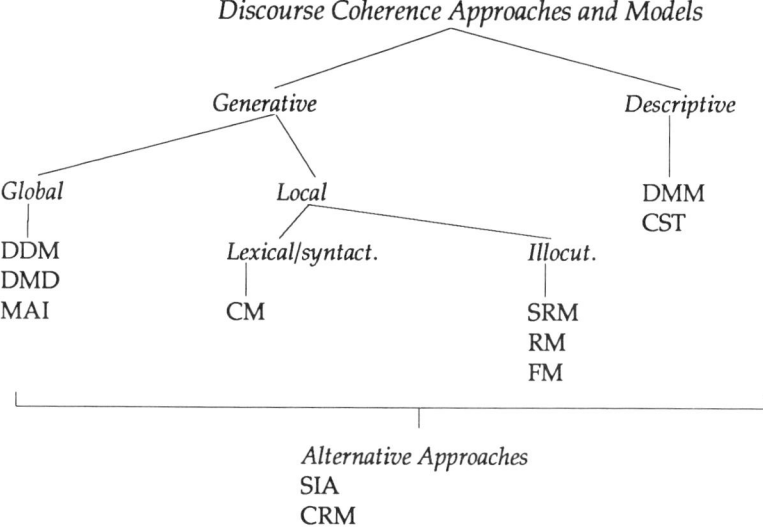

DDM	: Dynamic Discourse Model (Polanyi, 1985)
DMD	: Dynamic Models of Discourse (Sinclair, 1985)
MAI	: Models based on Artificial Intelligence (Hobbs, 1983)
CM	: Cohesion Model (Halliday & Hasan, 1976)
SRM	: Sequential Rules Model (Searle, 1969; Sachs et al., 1974)
RM	: Rational Model (Jacobs & Jackson, 1983)
FM	: Felicity Models (Weijdema et al., 1982; Haft-Van Rees, 1989)
DMM	: Descriptive Move Model (Goldberg, 1983)
CST	: Context Space Theory (Reichman, 1978)
SIA	: Situation-Interpreter based Approach (Hymes, 1974; Sigman, 1983)
CRM	: Cognitively Grounded Rules Model (Tracy, 1985)

2.3.1.   Global Models of Discourse Coherence

Global Models of Discourse Coherence, most notably associated with text and story grammars (Rumelhart, 1975; Van Dijk, 1977), artificial intelligence and cognitive science (Hobbs, 1979, 1983; Polanyi, 1985), suggest that well-formed discourse is constructed on the basis of general conversational or textual structures.

Global Models of Discourse Coherence were designed for analysing highly structured text-samples. These texts are used in restricted situations without the possible interventions of a conversational partner (e.g. compilers for programming languages) and events occurring in the nonlinguistic situation.

These models played a marginal role in preparing the IACV discourse framework and the current discourse coherence study.

Important IACV related issues in these models were:
- the way in which discourse coherence is defined in dependent and independent variables with mutually exclusive categories
- the use of variables to indicate discourse coherence at micro level (verbal and nonverbal act, moves, communication acts) and at macro level (turns, themes, contextual relations with preceding discourse).

Three major examples of Global Models are discussed below.

- *Dynamic Discourse Model (DDM)*

Polanyi (1985) discusses a 'Dynamic Discourse Model' in which structural properties are defined as so called 'discourse constituent units'. Discourse constituent units comprise sequences of units including stories, adjacency pairs and repairs. The units are seen as the building blocks of conversation. There are several subtypes of discourse constituent units, such as topic chain units, listing units and elaboration units. The Dynamic Discourse Model functions as a 'discourse parser theory'. A discourse consists of clauses encoding single propositions. Clauses form the input into the discourse parser. The Dynamic Discourse Model is an extreme example of generative discourse theories which describe discourse as a prestructured, generative process based on structural and semantic relations above the level of the sentence. The Dynamic Discourse Model was developed for highly constrained discourse situations.

- *Dynamic Models of Discourse (DMD)*

Sinclair (1985) proposes a model similar to the Dynamic Discourse Model, but less restrictive and rule governed. His Dynamic Model of Discourse provides a descriptive system to bring out the underlying similarities of structure in all discourse. The basic unit in Dynamic Models of Discourse is the 'move' (verbal or nonverbal acts with a specific function, such as initiations, responses, comments; see section 4.2.3.). Any utterance can follow any utterance. Although conversational partners tend to follow conventions in the organisation of social behaviour, there are no absolute rules, because people make mistakes or make use of conventions for indicating specific intentions (e.g. irony, anger). Each utterance sets the scene for the next. Sinclair's Dynamic Model of Discourse remains rather vague, compared with the Dynamic Discourse Model of Polanyi (1985).

- *Coherence Models based on Artificial Intelligence (AI)*

Coherence in spoken and written discourse is a growing area of interest in Artificial Intelligence (AI). Because of the importance of rule governed behaviour in AI, discourse coherence studies in AI can be seen as part of the Generative Global Model Approach.

According to Hobbs (1983), studies in AI, especially in inference systems, now allow the construction of a theory of discourse coherence for the representation and use of knowledge. Hobbs characterises discourse coherence in terms of a binary set of 'coherence relations' between segments of a discourse (a current sentence and the preceding discourse). He specifies four kinds of discourse coherence relations (temporal relations, evaluation relations, linkage relations and expansion relations). These relations are defined with computable precision in the framework of the inference component of a language processor. The typical inference system (embodied in a working computer programme) has four aspects: data, representation, operations, and control. 'Data' refers to the knowledge available to the system. In a natural language processing system an enormous amount of world knowledge must be accessed in understanding the most ordinary texts. 'Representation' refers to the formats in which the knowledge is stored. 'Operations' refers to the procedures which operate on the data represented. 'Control' refers to the choice of the operations applied and the order in which they are applied. These aspects seem to be inseparable in an AI theory of language use. Hobbs' theory mainly focuses on the operations that recognise discourse coherence. These operations work by attempting to construct chains of inference.

2.3.2. *Local Models of Discourse Coherence*

Local Models of Discourse Coherence account for what can, at any given point in discourse, follow what. These models describe the lexico-syntactic and illocutionary features that tie each verbal or nonverbal act to its immediate predecessor (Goldberg, 1983).

Studies on Local Models of Discourse Coherence were of more importance in preparing the discourse framework for IACV than the Global Models of Discourse Coherence. The studies on Local Models provided:
- adequate definitions for cohesion and coherence
- grammatical and lexico-semantic categories for coding cohesion (e.g. ellipsis, substitution, elaboration, lexical coreference)
- coding categories for communicative functions of verbal and nonverbal acts
- directions for segmenting a text into meaningful and functional entities.

We will discuss four major representatives of Local Models of Discourse Coherence.

2.3.2.1. Lexico-syntactic Models
- *The Cohesion Model*

The work of Halliday and Hasan (1976) on cohesion greatly differs from the Global Models of Discourse Coherence stated above. They catalogue a host of lexico-syntactic devices (e.g. pronominal reference, ellipsis, lexical coreference) used to create cohesion in discourse. Of particular interest is their idea that cohesion is a matter of degree. Thus for any given text a cohesiveness score may be computed by counting the cohesive devices present, or in their terminology, 'ties'. A tie is defined as a single instance of cohesion or the occurrence of one pair of cohesively related items (Halliday & Hasan, 1976,3). Freedle and Fine (1983) used an adapted version of Halliday and Hasan's coding of the degree of cohesion in a text. A series of oral story recalling tasks was performed with children aged from five to nine years old.

The processing difficulty of the used story texts was given in quantity and variety of occurred communication breakdowns, in terms of disruptions (e.g. hesitations, filled pauses, false starts). The results of their study indicate that a better performance (story recalling) of texts with a high cohesiveness score leads to a decrease in communication breakdowns.

Halliday and Hasan (1976,5) explain language as a multiple coding system comprising three levels of coding, or 'strata': the semantic (meaning), the lexico-grammatical (forms) and the phonological and orthographic (expressions). The more general meanings are expressed through the grammar, the more specific meanings through the vocabulary. Cohesion is expressed partly through the grammar (e.g. pronominal reference, substitution, ellipsis) and partly through the vocabulary (e.g. the use of general nouns, reiteration) Halliday and Hasan's primary concern is with written texts, produced by a single individual or seen from the view of a single conversational participant, yet their approach is claimed to be applicable to conversation (Freedle & Fine, 1983; Liles, 1985; Hartveldt, 1987).

Grammatical or lexical cohesion is not a prerequisite for discourse coherence. However, events occurring in the nonlinguistic situation, information presented through the preceding text, or linguistic context and the shared knowledge of conversational partners often lead to sequences of acts which have no cohesive devices and still appear relevant. Consider the following example:*

---

*All examples in this book are segments of original IACV transcripts.

(1)   Example of discourse coherence without cohesion

<a. asks c. to go home, because the play session was about to end>

a. ga je mee naar huis ?
    'are you going home with me ?'
c. pappa is honger
    'daddy is hungry'

---

a.   : *adult/caregiver*
c.   : *child*
'...'   : *paraphrase in English*

The reason for going home is probably that daddy will come home and be hungry. The video session ended at five o'clock in the afternoon. Before the session started the mother probably stated that daddy will be home and that they should hurry, because daddy will be hungry. According to Halliday and Hasan's model no cohesive devices have been used. While it is undoubtedly true that cohesion devices are usually present in the consecutive utterances and usually help to create a sense of connectedness, it will be necessary to consider more than just the discourse text. This aspect is discussed in section 2.5.1. concerning Situation-Interpreter based Models of Discourse Coherence.

2.3.2.2. Illocutionary Models of Discourse Coherence

Models of Illocutionary Discourse Coherence at the local level are based on speech act theory (Searle, 1969, Weijdema et al., 1982) or adjacency pair studies (Schegloff & Sacks, 1973). The cohesive connections which link turns are formalised in such a way that the range of possible and acceptable subsequent actions and speech acts are constrained. The type of data belonging to these models is best indicated as goal-directed and structured interpersonal interactions (e.g. interviews, problem-solving tasks).

Cohesion Models of this kind are primarily based upon the theoretical assumption that illocutionary acts or activity types are consistently and uniquely related to one another. Three important Illocutionary Models of Discourse Coherence are discussed below.

- *Sequencing Rules Models (SRM)*

Sequencing Rules Models attempt to model the orderliness of dialogue on the assumptions that utterances may be analysed as speech act categories and that the succession of utterances in conversation is regulated by rules that specify the range of speech act types that may appropriately follow any given speech act (Searle, 1969; Sacks et al., 1978).

These rules do not exist at the surface level of the discourse but at the level of the act performed. The centre-piece of the Sequencing Rules Models is the concept of the adjacency pair (two turn units). An adjacency pair consists of two parts, the first and the second turn. Utterances of the second turn are expected or conditionally relevant (Schegloff, 1972) to the first turn. If the expected second part does not follow the first part, the sequence will be deviant. Sequencing Rules Models identify constraints on what acts may follow another. These models are criticised on three points (Jacobs & Jackson, 1983).
1. Sequencing Rules Models offer predictions that do not hold up; conversational data do not demonstrate the level of regularity and predictability proposed (e.g. questions do not necessarily have answers but may be followed by questions or disagreeing responses). Adjacency pairs are not sufficient as a descriptive unit in discourse. It is often the case that other verbal or nonverbal act items are inserted. Consider the following example:

(2)   *Example of embedded/inserted sequence*

<a. is looking for a book>

a.	waar is het boekje, Linda?		Q-1
	'where is the book, Linda?'		
c.	welk boek?	Q-2	
	'which book?'		
a.	het plaatjesboek	A-2	
	'the picture-book'		
c.	Oh: daar op de vloer		A-1
	'Oh: there on the floor'		

Q.   : Question
A.   : Answer
a.   : adult/caregiver
c.   : child
'...' : Paraphrase in English

In example (2) a question(Q)-answer(A) sequence has been inserted between the initial question (Q-1) and the final answer (A-1). As far as the structure concerns, the adjacency-pair is a descriptive unit unsatisfactory on at least two points (McTear, 1985). The adjacency-pair does not account for relations between all the possible utterances and nonverbal acts in conversation and the units of the adjacency-pair are not defined explicitly enough in order to delimit the range of possible second pair parts and thus to distinguish between possible and impossible responses.

2. Sequencing Rules Models are difficult to use in finding an account for which speech acts can and cannot initiate an adjacency-pair. A large number of speech acts establish no obvious conditional relevance. Statements, for example, do not place any obvious restrictions on how to make a reply, if they require a reply at all.
3. Sequencing Rules Models provide no basis for explaining why a rule takes the form it does. This also means that there is no way to generate other rules of conversation.

- *Rational Model Approach (RM)*

A so-called Rational Model is presented as an alternative for the Sequencing Rules Models by Jacobs and Jackson (1983). It is important to recognise that in Sequencing Rules Models speech acts are used as the structural units upon which coherent sequential relations are defined. The functional properties of speech acts do not enter into such definitions. A Rational Model of conversational coherence must show how coherent dialogue results from the application of practical reasoning to conventionally defined means of achieving goals (e.g. speech acts). A Rational Model must include systems of rules defining appropriate means of achieving particular kinds of goals, such as rules and preconditions for performing requests, where the goal is to let an addressee perform some future act. According to the Rational Model the coherent discourse must be considered as the orderly output of practical reasoning about goals, constrained by institutionally defined means of achieving those goals.

Jacobs and Jackson compare conversational structure with formal game or play structure. In their view coherent dialogue is based on two manifest properties:
- goal orientation (intrinsic or extrinsic to the conversation)
- alignment (conversational partners try to achieve a working agreement).

A major problem of the Rational Model is its general, philosophical presentation; it is difficult to apply the Rational Model as a descriptive and explanatory model. No adequate Rational Model yet exists. Only in the Cognitively Grounded Rules Model (Tracy, 1985) some central notions of the Rational Model are applied (section 2.5.2).

- *Felicity Models (FM)*

These models are based on a hierarchical organisation of the conversation consisting of (sub)sequences and preconditions for successful communicative interaction (e.g. the sincerity conditions defined by Searle, 1969). The subsequent utterances form a topic that can be interrupted by speech acts not belonging to the current topic, but are used for modifying the topic (e.g. the so-called 'side-sequences' in Jefferson, 1972). The modifications in a current topic are in fact expansions of sequential (or adjacency) pairs. There are several possibilities for expanding sequential (or adjacency) pairs in using one or more pre-sequences, one or more embedded sequences and one or more post-sequences (Haft-Van Rees, 1989). Speakers use these expansions to ensure an unambiguous and successful conversation. Felicity Models of Discourse Coherence describe the underlying preconditions of conversational partners using these kind of expansions, and more specifically the speech acts to fulfil the preconditions. The problem is to determine which preconditions are needed for successful communication. Haft-Van Rees (1989) discusses absolute and relative preconditions. Absolute preconditions are obligatory conditions for all kinds of spoken discourse (e.g. the intelligibility of spoken utterances). Relative preconditions are only of relevance in specific situations; they are only activated when needed.

Weijdema et al. (1982) present an outline for describing discourse coherence with the use of preconditions and related speech acts. The main part of their model is derived from Searle (1969) with a major adaptation in defining the speech act. According to Searle (1969) speech acts are primarily acts performed by the speaker. In conversations language is used actively by both partners, speaker and hearer. The speaker as well as the listener are bound to preconditions concerning successful interaction. Dependent upon the relevant preconditions in interactions, speakers and listeners use different speech acts. The hierarchical architecture of discourse as presupposed presented by Weijdema et al. (1982) offers good opportunities to study the rationale behind the use of speech acts in conversations. The difficulty with their approach is its applicability to studies of spoken discourse consisting of casual exchanges between familiar interactants. The reason of the weakness of the system is that these casual exchanges are characterised by a great number of different discourse structures and organisational patterns which the model does not explain accurately.

2.4. Descriptive Models of Discourse Coherence

Descriptive Models examine discourse for recurring patterns without claiming that these patterns will always occur. Descriptive Models stress that conversations are managed at the level of utterances (Goldberg, 1983, Reichman, 1978; Keenan & Schieffelin, 1976). The description focuses on conversational moves (in Goldberg's Move Model, 1983) or on particular lexico-syntactic cohesive links which function as potential transitions for topic shift and speaker orientation (in Reichman's Context Space Model, 1978).

The Descriptive Models of Discourse Coherence indicate the importance of specifying the origins of coherence relations at the level of the utterance.

The information from these Descriptive Models was explicitly used in the set-up of the IACV pragmatic analysis procedures, especially in the definition of several initiation types (initiations which introduce a new theme, initiations which reintroduce a theme and initiations which elaborate on an immediately preceding theme, see section 4.3.1.) and in incorporating coding categories for discourse coherence and communicative functions at micro level (at verbal and nonverbal act level, see section 4.3.7.) as well as on macro level (at the level of conversational turns and subsequences, see the sections 4.2.4. and 4.2.5.). Two representatives of these Descriptive Models will be discussed.

- *Descriptive Move Model (DMM)*

Within a Descriptive Move Model utterances are categorised as moves in terms of their lexico-semantic ties with preceding utterances both within and between turns. The Descriptive Move Model distinguishes four types of moves; introducing, reintroducing, progressive-holding and holding moves. Goldberg describes the Descriptive Move Model as a kind of micro analysis which makes it possible to identify thematic and sequential boundaries at the point at which they occur. This seems especially useful in cases of subtle topic shift for determining where (and why) one topic or sequence ended and the next began. When the Descriptive Move Model is applied to natural discourse, clear sequencing preferences are not obvious.

There appear to be no absolute constraints on what type of move may follow another. According to Tracy (1985) Goldberg's failure to find absolute constraints may be related to the units of analysis used. The units in Descriptive Move Model probably are too gross to capture the constraints that exist. More likely, however, is the possibility that there are no absolute constraints. Sequencing Rules Models (discussed in section 2.3.2.2) emphasise regularity; the Descriptive Move Model emphasises the lack of regularity.

- *Context Space Theory*

Reichman (1978) discusses a model of studying cohesion, based on so-called 'context spaces'. A 'context space' is defined as a group of utterances referring to a single issue or episode. Reichman defines a conversation as a sequence of utterances; at a deeper level it is a structured entity of which utterances can be parsed into hierarchically ordered context spaces. Conversational coherence depends upon a lack of conflict between respective discourse models of the participants. Conflicts are prevented by speakers following high level syntactic and semantic relational rules that enable listeners to identify their discourse models.

2.5. Alternatives to Generative and Descriptive Models

The major drawback of Generative Models (section 2.3.) is that they do not specify how, within the discourse modelled, each utterance or nonverbal act is formally and specifically related to the one preceding it. The structure of discourse is already determined. The spoken text has to be performed in the order specified by the discourse type (Goldberg, 1983; Tracy, 1985; Haft-Van Rees, 1989). For example the structure of formal conversations between a doctor and his patient, telephone conversations, buying and selling routines and content specific features in joke-telling.

The Global Models (section 3.3.1.) of discourse work best as discourse comprehension models, rather than encoding ones. Illocutionary Cohesion Models (section 2.3.2.2.) are based upon the assumption that illocutionary acts are consistently and uniquely related to one another. These relations are evoked by sequencing rules and constraints, proposed by the theoretical model. These rules are supposed to generate well-formed coherence pairs and conventionalised discourse units (e.g. the side-sequences and repairs described by Jefferson, 1972).

In fact, these sequencing rules reflect statistical probabilities of two acts following each other. So, all that can be claimed is that, according to the model, there is a certain chance of sequential occurrence of two speech acts. The Generative Models discussed are useful for ideal and prestructured discourse situations in which predictions about sequential order of acts can be made. However, examination of most discourses indicates that communicative interaction is rarely prestructured, but usually reveals a blend of diverse organisational discourse structures (e.g. parallel talk, unexpected topic shifts, faulty responses, consecutive initiations). The development of a conversation rarely follows one predictive pattern throughout. The most serious problem with the alternative for Generative Models, the Descriptive Models (section 2.4.), is the

lack of regularity. Goldberg's Move Model does not specify constraints on what type of move may follow another. In the discussions of the models reviewed it seems undoubtedly true that discourse coherence is usually present between and within consecutive utterances and that we need to consider more than just the discourse text. Tracy (1985) discusses two other factors influencing discourse coherence:
- the conversational situation or context and
- the mind (knowledge) of the listener, interpreter.

She presents two alternative approaches that can be seen as reactions to and criticisms of generative and descriptive approaches to conversational coherence, rather than independent explanatory positions. These two alternatives are discussed in the next sections.

2.5.1. Situation-Interpreter based Approaches

Situation-Interpreter Approaches (Hymes, 1974; Sigman, 1983) try to explain discourse coherence by using information of the situational and relational knowledge expressed by the conversational partners. Consider the following example:

(3)   *Coherence because of shared situational knowledge*

    <c. shows her 'medical kit' (toy) to a.>

    a. zo een heb je ook van oma gekregen
       'grandmother gave you a similar toy'
    c. pappa en jij zijn met vakantie geweest
       'you and daddy were on holiday'

a.	: *adult/caregiver*
c.	: *child*
'...'	: *Paraphrases in English*

In this example the sequence seems connected, not because of the specifics about the discourse but because child and caregiver share the same situational knowledge. During the time that her parents were on holiday, the child was with her grandmother. Grandmother gave her grandchild a present. The caregiver remembers that situation while playing with the child during the video session. So, the relevance of one utterance to another then depends on the preceding linguistic context. Situation-Interpreter Approaches indicate two problems with Generative and Descriptive Models:

- the variability of situation and
- the reactions of the conversational partners.

An adequate theory of conversational coherence will need to take into account the role of the situation. For that reason transcripts of conversations need fully specified situational information.

2.5.2. *Cognitively Grounded Conversational Rules Model*

Tracy (1985) presents a model for studying discourse coherence based on two key parts: a 'Rules Approach' and a 'Cognitively Grounded Approach'.
The Rules Approach to conversational sequencing specifies:
- preferred types of conversational extensions
- preferred response types
- preferred manners for responding
- conditions under which the preference operates.

The Cognitively Grounded Approach to conversational rules identifies the regularities that exist in the sequencing of conversational units and communicator judgements of appropriateness, and explains (on cognitively motivated grounds) why these regularities occur. Communication is a complex, cognitively based task which taxes the mental resources and capacities of the conversants. Conversational extensions which draw upon typical inference patterns will be preferred to those that draw upon atypical ones. Tracy's model shows some similarity to the Rational Model Approach discussed above (section 2.3.2.2), in which rational rules are formulated by communicators and are used to pursue goals in conversation.

The Situation-Interpreter based Approaches and the Cognitively Grounded Conversational Rules Model indicated the need for fully specified transcripts and the incorporation of coding categories for appropriateness and correctness of verbal and nonverbal acts in the IACV pragmatic analysis procedure. Judgements of appropriate and correct language behaviour were made from the point of view of the conversational partners.

2.6. Summary

In this chapter several existing approaches and models for the study of discourse coherence were discussed briefly. The survey presented did not intend to offer a complete and fully specified overview of approaches and models.

Nevertheless the three main groups of models and approaches discussed (Generative, Descriptive and Alternative Models and Approaches) gave a general impression of the vast and different number of studies conducted in the field of discourse coherence. Each main section in this chapter indicated the aspects of the models discussed which are relevant to the current study. Table 2.1. on the next two pages summarises the main linguistic means of the models reviewed and also indicates the issues of these models which are relevant to the study at issue. The discourse coherence studies in general offered an important group of linguistic categories and also presented background knowledge for establishing a framework for the study of discourse coherence in conversations between caregivers and their SLI or NLA children. Specific information about the research design and the definitions of linguistic categories used are discussed in chapter 4. Chapter 3 presents a survey of discourse (coherence) studies in caregiver-child conversations. The information in chapter 3 leads to the specification of the research hypotheses, to be formulated in chapter 4. The linguistic categories introduced and the definitions given in chapter 4 constitute the instrument for finding (counter-) evidence for the hypotheses formulated.

TABLE 2.1. Coherency Models and their relevance to IACV

Coherency Model	Linguistic Means	Relevant Issues to IACV*
Dynamic Discourse Model (DDM) (Polanyi, 1985)	Discourse parser theory; building blocks (dcu's), defined on clause level	Strictly rule-governed; no direct application in IACV study
Dynamic Model of Discourse (DMD) (Sinclair, 1985)	Basic unit: sentence or move. Combination of different linguistic means: speech-acts, reference-types, ellipsis	Descriptive and dynamic approach. Combination of linguistic means at micro and macro level, using morpho-syntactical and pragmatical categories. Hierarchical structuring of discourse from the point of view of the partners
Artificial Intelligence (AI) Models (Hobbs, 1983)	Inference systems. Unit: coherence relations (e.g. conjunctives)	Mutual exclusiveness. Interpretation of current acts in light of the preceding discourse
Cohesion Model (CM) (Halliday & Hasan, 1976)	Cohesive ties in terms of grammatical and lexical categories for verbal acts	Definition of IACV variables (feedback, imitation, ellipsis, correctness of language use)
Sequencing Rules Model (SRM) and (Sacks et al, 1978)	Sequences of speech acts and adjacency-pairs	Definition of co-occurring speech acts in various communication functions, the definition of turns
Rational Model (RM) Jacobs & Jackson, 1983) *and Felicity Models (FM)* Weijdema et al, 1982)	Preconditions (relative and absolute) for speech act occurrence	Definition of use-incorrect acts, clarification request, communication breakdowns
Descriptive Move Model (DMM) (Goldberg, 1985)	Utterances which are characterised as moves; thematical organisation	Definition of initiatives related to thematical interpretation of participants

* Definitions of the IACV variables mentioned in this column are given in chapter 4.

TABLE 2.1. (Continued)

Coherency Model	Linguistic Means	Relevant Issues to IACV*
Context Space Theory (CST) (Reichman, 1978; Sanders, 1983)	Context space; a group of utterances that refer to an issue/episode	Definition of context, situation and topic. Themes as structural and functional entities of discourse organisation
Situation-Interpreter Based Approach (SIA) (Hymes, 1974)	Decisive role of the situation and shared knowledge of the partners in judging coherency	The format of the transcript and the kind and amount of context information Definition of incorrectness from the interpreter's viewpoint. Only information in the discourse, situation and context itself can be used for interpreting coherence
Cognitively Grounded Rules Model (CRM) (Tracy, 1985)	Topical structuring of discourse and the role of context, linkage strategies	Theme-topic structure. Importance of context and situation. Distinguishing three kinds of initiatives, responses and comments Coherence studied from the perspective of both partners

* Definitions of the IACV variables mentioned in this column are given in chapter 4.

CHAPTER 3

Coherence in adult–child conversations

As mentioned in the introductory chapter, the main focus of this book is on how children initiate and sustain conversations with their primary caretakers. The research published on this topic is surprisingly limited. This chapter presents an overview of the literature on discourse coherence in conversations of NLA and SLI children with their caregivers. The scope of this study does not permit a complete description of the research conducted in the area of normal and deviant language behaviour of children at preschool ages. Bloom and Lahey (1978), Wanner and Gleitman (1982), Gallagher and Prutting (1983), Wells (1981, 1985), McTear (1985), Franklin and Barten (1988) and Kessel (1988) present good overviews.

3.1. Coherence: the shaping of discourse

One of the ultimate goals of language acquisition is the development of abilities to obtain information from observed situational events and perceived utterances, to relate that information to existing 'knowledge of the world' (meaning relations), and to prepare contingent messages with respect to shared or given knowledge of the conversational partners. Perhaps the most salient feature of coherent discourse and the one that attests most strongly to sequences of connected information, is the prevalence of (verbal and nonverbal) acts or pairs of acts which are structurally and functionally related to each other (e.g. question-answer, request-comply, inform-acknowledge). This relation is expressed in 'turn exchanges' (the taking of initiations, responses, comments) of the conversational partners (Wells, MacLure & Montgomery, 1981; Barnes, Gutfreund & Satterly, 1983). The exchange of turns has been regarded as a central component of conversation (Sacks et al., 1974; Duncan & Fiske, 1977).

The results of many studies dealing with adult conversation indicate that cooperative conversation not only requires knowledge of the linguistic code but also know-how about nonlinguistic and social codes (DeMaio, 1982). The child has to learn these codes by inferring rules and conventions for turn-taking, turn-giving and maintenance from contingent discourse with his caregiver (Sander, 1977; Ervin-Tripp, 1979). Knowledge about turn-exchanges is necessary for learning how to organise and to structure meaning relations in a cohesive way during conversation (Sacks et al., 1974). It is important to realise that conversation is primordial because the cradle of all language use is the conversational turn-taking between child and caregiver (Bruner, 1983). The origin of conversational behaviour is found in early infancy in the prelinguistic stage, that is before children start talking (Bullowa, 1979). The early interactions between mother and child are mostly referred to as 'proto-conversations' (Bateson, 1975). Detailed analyses of such interactions have shown that they have the appearance of mature conversations, like turn-taking with minimal overlap, performance of turn-like exchanges based on mostly simple initiative-response sequences. These proto-conversations seem to function as the basis for the subsequent development of communicative abilities. It is nowadays widely accepted that mothers (and other caregivers) adapt their behaviour when interacting with babies and young children. The majority of research reports has been focused on mothers' linguistic adaptations and their potential significance for the child's language development, called 'Baby-Talk' (Snow & Ferguson, 1977) or 'Child Directed Speech' (Snow, 1986). Caregivers also exhibit other adapted (nonverbal) behaviours in their interaction with babies and young children (e.g. facial expressions, frowning, extended eye-contact, vocalisations). In that case the term 'Child Adjusted Register (CAR)' (Dromi, 1989) is preferred to 'Baby Talk' and 'Child Directed Speech'. It is suggested that CAR's prime function is to assist in the initiation, regulation and structuring of interaction with immature interactional partners.

Mothers often synchronise their turns with what the child is doing or is likely to do next in such a way that the child's actions are seen as responses. Consider the following example from Rachel and her mother.

(4)   *Example of proto-conversation NLA child (0;3) and her mother*

<c. lies on chair in living-room and is crying and
a. kneels down before c.>

1. a.   vind je het niet meer leuk?
        'don't you like this anymore?,'
2. a.   //nee hè?
        '//you don't, do you?'
3. a.   pakt c. op armen] en staat op
        'takes c. in her arms] and gets up'
4. a.   zie je dit is beter hè?
        <c. stopt met huilen>
        'you see, this is better isn't it?'
        '<c. stops crying>'

---

a.   : *adult/caregiver*
c.   : *child*
<...>   : *contextual and situational information*
//....] : *start (//) and end (]) of overlap of acts*
'...'   : *paraphrase in English*

The first verbal act (1) functions as an initiative to which 'yes' or 'no' are appropriate responses. The crying of Rachel is interpreted as a confirming response of dissatisfaction. Rachel's response (stops crying) to the changing situation (moving from the chair to the arms of the mother) is interpreted as a child response.

The role of mothers and other caregivers in structuring conversational exchanges involving their infants extends to later stages. The language used by adults for this purpose includes interrogatives (subject-auxiliary inversions and interrogatives by intonation only), tag questions (declaratives + tag (= hè?)), role- play in which complete dialogues are presented by the caregiver as exemplified frameworks of turn-exchanges (e.g. mother-father role-play, simulating telephone conversations, simulating doctor-patient conversations). Frequently it is the child who initiates discourse while the mother replies (Schachter, 1979).

The structure and function of alternating moves assist the child in identifying strategies for sustaining interaction (e.g. the use of pronominal reference, acknowledging (verbal and nonverbal), imitations). Coherence is a product of the mutual process of producing and interpreting verbal and nonverbal acts in discourse. The adult assists the child in his interpretation of communicative messages, by using the latter's cues indicating coherency, but still the child (as hearer or observer) has to interpret these as such. Gradually the child acquires more linguistic and lexical strategies; his comprehension and performance of cohesive devices in discourse increase.

The study at hand is set up in order to investigate differences in creating discourse coherence across verbal and nonverbal acts in conversations between caregivers and their SLI or NLA children.

The following example shows Noortje (an NLA child at the age of 3;0) engaged in discourse with her mother. Mother and child are acting contingently in the use of initiations, responses and comments intentioned to maintain the conversational topic. The child mostly responds to her mother's questions and remarks. The framework of discourse (the use of pronominal reference, requests for clarification and confirmatives) is built up by the caregiver. Similar examples are given in Bruner (1975), Bloom et al. (1976), Cross (1978), Barnes (1983) and Foster (1986). The behaviours of child and adult are finely tuned to each other in a mutual process of information exchange in order to establish coherent discourse.

(5)   *Fragment of coherent discourse between NLA child (3;0) and her mother*

<adult and child are playing with stove)

    a. //wat ben je aan het koken?
        '//what are you cooking?'
    a. kijkt naar activiteiten c.]
        'watches child's activities]'
    c. //soep -- turn-exchange
        <kijkt in pan op fornuisje/a. kijkt c.>
        '//soup'
        '<c. looks into pan on stove/ a. watches c.>'
    c. roert met lepel in pan]
        'stirs with spoon in pan]'
    a. (.2) //kijkt in pan -- turn-exchange
        <op fornuis>
        (.2) '//looks into the pan'
        '<on stove>'

a. hm wat ruikt dat lekker zeg]
'hm: this does smell good, doesn't it?]'
a. kijkt naar activiteiten van c.
   <c. roert verder in pan>
   'watches activities of c.'
   '<c. keeps on stirring in pan>'
a. dat is tomatensoep hè?
   'that's tomato soup, isn't it?'
c. //verzet pan naar ander kookpunt ---------------------- *turn-exchange*
   '//moves pan to another burner'
c. nee: sampijonne soep is dat]
   'no, that's mushroom soup]'
a. och ja wat stom --- *turn-exchange*
   'oh, how stupid of me'
a. dat eet je toch ook graag hè?
   'you like it though, don't you?'
c. ja --- *turn-exchange*
   'yes'
a. ik ga maar al vast de tafel dekken ------------------------ *turn-exchange*
   'I am going to lay the table'

---

a.   : *adult/caregiver*
c.   : *child*
<...>   : *contextual and situational information*
//....] : *start (//) and end (]) of overlap of consecutive acts*
'...'   : *paraphrase in English*
(.2)   : *2 seconds of interval-time between two verbal acts*
-----   *turn-exchange* : *defined in section 4.2.4*

3.2. Current status of interaction studies with Specific Language Impaired (SLI) children

Only recently discourse organisation has evolved as a topic to be studied in the context of language and communication development. This evolution is predominantly caused by the shift in emphasis from sentence structure (Chomsky, 1965; McNeill, 1966) to relationships between structure and meaning (Bloom, 1970; Brown, 1973; Bloom et al, 1980) and the subsequent emergence of functionally based grammars in language acquisition research (Bates, 1976; Karmiloff-Smith, 1979). Functional analyses have focused attention on single utterances, on the influence of the context on conversational turn exchanges (Sacks et al, 1974; Halliday, 1975; Sander, 1977; Ervin-Tripp, 1979; DeMaio, 1982), on topic manipulation (Keenan & Schieffelin, 1976; Fine, 1978; Brinton & Fujiki, 1984; Foster, 1986; Luszcz, 1983) and on 'Baby Talk' (Snow & Ferguson, 1977; Gleitman, Newport & Gleitman, 1984; Hirsch-Pasek & Treiman, 1984; Snow, 1986) or 'Child Adjusted Register' (Dromi, 1989, Conti-Ramsden, 1990).

Simultaneously, arguments from related studies in general and from what is similar in children' acquisition across different languages have increasingly shaped our thoughts about language acquisition (Chomsky, 1965; Bowerman, 1973; Ferguson, 1977) and cross-linguistic, cross-cultural studies (Slobin, 1982).

This increase of research interest has also influenced the study of characteristics found in the communication of SLI children (Lasky & Klopp, 1983; Fey & Leonard, 1983; Conti-Ramsden, 1987, 1989). Although the lexical and syntactic abilities of SLI children are still a major concern, investigators have broadened their research area by including consideration of the ways SLI children use their language or practise their communicative abilities. The study of pragmatic skills among SLI children is still in an early stage of development.

Although research increased the amount of interaction studies with SLI children, many of these investigations are preliminary and descriptive in nature (Barnes et al., 1983).

Discussing the results of these investigations leads to a somewhat confusing picture. We therefore will present a discussion of five main research areas in which discourse coherence is studied, instead of giving a complete overview of studies in normal and deviant language development.

These five areas are: Turn-taking, Child Adjusted Register (CAR), Topic and Theme Management, Communication Acts, and Communication Breakdowns and Repairs.

3.2.1. *Turn-taking*

Conversational participation comprises a rather broad category of features, which includes a number of sociolinguistic skills. Studies in this area focus attention on the reciprocal nature of conversation and the use of skills for initiation, elaboration or termination of conversational themes (e.g. children's role as speaker-initiator and as listener-respondent).

The research on conversational turn-taking by children has emerged from responses to two traditional positions regarding the nature of language acquisition and language use. The first position, introduced by Chomsky (1965), stated that children have an innate capacity to develop language. Because of the many irregularities, disfluencies and grammatical inconsistencies in adult talk to children, children cannot derive underlying rules from these language data. They have to rely on an innate language acquisition device. The work of Chomsky (1965) has led to an increased interest in studying the role of environmental input into children from the natural interaction with their mothers and other adults. The second position, introduced even earlier by Piaget (1959; first published in French in 1926), stated that children's speaking to peers is egocentric and that without adult guidance children are unable to maintain conversation.

Piaget's findings are based on studies of children's conversations in a nursery school in Geneva (Switzerland) and also on experiments designed to test children's abilities to communicate information effectively. Piaget came to the conclusion that children's early communication is deficient and he explained this deficiency in terms of a theory on child egocentrism. Piaget's studies have led to experimental investigations in referential communication, in which it was suggested that the young children's egocentrism caused an inability to take the listener's perspective into account. The results of these studies were in conflict with this suggestion (for discussion and overview see Ginsburg & Opper, 1979; Seibert & Hogan, 1984). The fact that children seem to consider the conversational perspective of the other partner in conversations caused an increased interest in examining turn-exchanges among children and their caregivers (Snow, 1972, 1977; Bruner, 1975, 1978, 1983).

Bloom et al. (1976) observed that, by the end of the sensorimotor period, children begin to follow the basic rules of turn-taking. The children have learned that conversational participants take turns when talking and that individuals mostly start speaking as a consequence of others speaking to them. Relatively few studies have examined conversational turn-taking in mother-child dyads with children at chronological ages 2;6 to 4;0.

Kaye and Charney (1980) have described general turn-taking characteristics of mothers interacting in a free-play situation with their children at age 2;2 and at age 2;6. The majority of the discourses consisted of fluent turn-exchanges. Children were found to interrupt more frequently than the mothers. There were few overlaps (simultaneously occurring turns). Turn-exchanges were still primarily managed by the adult.

Most of the information about turn-exchanges in conversations with children at preschool age (from 3;0 to 4;0) specifically involved studies concerning overlaps (verbal and nonverbal acts) or interruptions (only verbal acts).

These studies were mostly based on peer interactions (Ervin-Tripp, 1979; Garvey & Berninger, 1981; McTear, 1985). From these studies it appears that more overlaps occur at an earlier age.

Shifts of topic within interruptions were also examined and found more likely to be ignored by younger children than by older children (Ervin-Tripp, 1979). Another interesting result was that the younger the child who interrupted, the more likely the adult was to ignore the interruptions. Although conversational turn-taking violations have been described in peer interactions between pre-school children, this behaviour has not been described as extensively in mother-child conversation (McTear, 1985). In the light of studies concerning Child Adjusted Register (CAR), turn-taking skills appeared to be an important aspect of the facilitative role adults play in their conversations with children. For example the adults' use of elicitations, imitations, content questions and clarification requests have a tutorial function for the child learning to judge the content and transition points in current turns. If mothers are facilitative in teaching the child to understand turn-transition points, one might expect that mothers more likely than their children adhere to the turn-taking rule in our cultural society that only one participant can talk at a time, and that they would be more likely to discontinue talking if such a violation occurs.

Bedrosian et al. (1988) studied thirty children (fifteen girls and fifteen boys, at pre-school age) and their mothers. Ten minute free-play sessions were videotaped through a one-way mirror. All utterances used in the dyadic interactions were orthographically transcribed and coded on general characteristics of turn-taking violations, parallel talk; partners start to talk at the same time, the partner who initiates an interruption, the kind of turn-taking repairs made (e.g. one of the partners stops talking and repeats his utterance, the use of interruption markers ("sorry", "again")), the way in which topic consistency is involved (e.g. introduction of a new topic, the maintenance of the current topic) and the communicative intention of the acts involved (e.g. informative acts, requests, refusals, acknowledgements).

Because mothers have been found to facilitate their children's development of discourse skills, the authors expected to find a similar facilitative role in relation to managing turn--taking violations (e.g. overlaps). If mothers were facilitative, one might expect that they would use fewer turn-taking violations than their children and that they would be more likely to exhibit repair mechanisms when such violations occur (Kaye & Charney, 1980, 1981). Contrary to these expectations the authors observed that mothers exhibited more turn-taking violations than did their children. The role of mothers, therefore, appeared to be more closely aligned with control rather than facilitation. In addition, mothers tended not to extend their interruptions beyond one conversational turn. In looking at the dyad as a unit, the frequency of overlaps decreased significantly with the children increasing in age. Mothers seemed to be more sensitive to their child's growing ability to engage in cooperative discourse. Mothers and children tended to maintain topics within overlaps and between consecutive overlaps.

The counter-evidence found (more turn-taking violations by the adult) in the study of Bedrosian et al. (1988) might be partially attributed to a more equal status between children than between mothers and children. This suggests that mothers are more sensitive to linguistic structures and meaning relations within utterances and ongoing topics than would be the case in normal adult-adult communication and in peer interactions between children.

It is difficult to compare these results with studies involving SLI children. There are only a few studies of relevance in this area.

According to several SLI studies reviewed by Fey and Leonard (1983), it seems that SLI children compared to NLA children of the same age are markedly deficient as initiators of social conversational interaction and mostly react only to elicitations.

As reported in the Fey and Leonard overview, most of the investigators noted that SLI children appeared overly reliant on the use of backchannelling (defined as an acknowledgement of the preceding act without the intention of taking over the conversational turn).

Some studies signal few incidences of backchannelling among SLI children (Watson, 1977; Fey, Leonard & Wilcox, 1981). These studies however do not suggest that the incidences of backchannel signalling in SLI children are similar to those of NLA children.

In the Watson (1977) study the incidence of backchannelling was found to be more frequent among SLI children than NLA children at comparable ages. Fey et al. (1981) reported less backchannelling employed by SLI children communicating with partners of the same age than younger (MLU matched) partners. These studies suggest that SLI children appear to rely more heavily on backchannelling as a means of avoiding a turn in conversations with adults and peers than NLA children. Several other studies in which the communicative interaction of caregivers and their NLA or SLI children is investigated, indicate that both groups of children differ slightly in the frequencies of interaction patterns used (such as imitations, responses, speech acts). Differences between NLA and SLI children were observed in the relationships between measures of linguistic maturity of NLA children and their caregivers' interaction patterns which were not apparent to SLI children and their caregivers. Caregivers' use of interaction patterns such as imitations, questions, acknowledgements, nonverbal acts, feedback and strategies for sustaining topic all were described as being related to language maturation measures (e.g. MLU, responsiveness, initiations, expansions) of the NLA child. These relationships were infrequent and much weaker with SLI children (DeMaio, 1982; Lasky &

Klopp, 1982; Conti-Ramsden & Friel-Patti, 1983; Roth & Spekman, 1984). Presumably SLI children are willing to participate in conversation, but do not have full access to a broad and necessary scope of linguistic skills. NLA and SLI children seem to learn their roles in conversation by imitating the language behaviour of their caregivers. NLA children, SLI children and both their caregivers show the same frequencies of interaction patterns. Problems in language and communication development of SLI children should therefore be sought for in differences in the relationships between caregivers' interaction patterns and indices of the children's language development of NLA and SLI children.

Bloom et al. (1976) observed that children first start with exact (or partial) imitations of immediately preceding adult utterances. From Stage 1 to Stage 5* they acquire a lot of information about possible functions of verbs. From Stage 2 they start to learn how to expand the main verb clause by adding new information (adverbial phrases). Children do this by gradually expanding a preceding utterance of the adult. When a topic is originated by a prior child utterance, the intervening adult utterance functions as a prompt in the form of an imitation, or asks for an expansion by the child. Imitation and elicitation of imitative behaviour play important roles in the learning of turn-taking. Besides the effect of learning grammatical and lexical functions of verbs and verbal relations, imitations also seem to indicate how to use these devices in order to create textual relations between clauses, utterances and turn-exchanges. Indirectly this leads to the use of referential language and the notion of contingency (see also Bloom et al., 1976). It is probably for that reason that mothers of SLI children continue to use a large number of repetitions of their own utterances (verbal imitations) in order to provide a framework for contingent interaction and verbal learning (Lasky & Klopp, 1982).

3.2.2. *Child Adjusted Register (CAR)*
According to several studies in adult-child communication the conversational participation of children depends on the social context (situation and partner) of communicative interaction (Fey et al., 1981; Leonard, 1986). Most of these studies indicate that caregivers simplify and adjust their communication to young SLI children in much the same way when talking to young NLA children.

McDonald and Pien (1982) have suggested that caregivers attempting to elicit children's conversational participation may do so by using short speaking turns and encouraging equal participation, whereas mothers wishing to

*Defined according to Brown (1973); Stage 1, MLU < 2.0; Stage 2, MLU 2.00–2.75 and Stage 5, MLU 3.5–4.0.

direct or control the child's behaviour are likely to use longer speaking turns and dominate the conversation.

Schodorf and Edwards (1983) have found evidence for some differences in the Child Adjusted Register (CAR) when parents are talking to their child (SLI and NLA, matched to MLU) in structured teaching tasks as opposed to free-play sessions. In teaching tasks parents of SLI children and parents of NLA children produced the same number of utterances addressed to the child. However, the utterances to SLI children were significantly shorter in length (MLU) than those spoken to NLA children. Nevertheless, during free-play sessions, parents of SLI children used utterances similar in MLU to those spoken by parents of NLA children. Yet they produced fewer utterances per turn than did the parents of NLA children. So, during the free-play sessions SLI children tend to receive a shorter adult language input (in terms of acts per turn), which is also more directly related to preceding acts and situation (Schodorf & Edwards, 1983; Conti-Ramsden, 1983, 1988, 1990). Mothers of SLI children used exact repetitions of their own utterances more often than mothers of NLA children (Lasky & Klopp, 1982). These repetitions probably serve as conversational repair strategies (e.g. the repetition of an utterance in order to repair a conversational breakdown caused by a misunderstanding or an ambiguous utterance) and also offer good opportunities to continue conversational participation (Bloom et al., 1976; Wells, 1981).

The characteristics of CAR mentioned above seem to be contradictory to studies concerning the 'Motherese Hypothesis'. This hypothesis proclaims that the simplest input speeds up language learning best (Snow, 1972; Cross, 1977). Furrow et al. (1979) have found that the largest progress in early language learning was made by NLA children (at age 1;6/MLU 1.00) who received the simplest input in terms of maternal utterance length (MLU and mean number of words per utterance).

This implies that SLI children compared to young NLA children (age 1;6) and receiving a simpler CAR, should be similar in their progress of language learning to NLA children. To generalise, such a relationship is counter intuitive (Gleitman, Newport & Gleitman, 1984). The Furrow study must certainly be evaluated in relation to the small, socially homogeneous sample they studied.

Gleitman et al. (1984) indicate serious problems in studies concerning the influence of adult language input and child language progress. Most studies are based on rather small populations, are cross-sectional and use relatively long periods between subsequent language learning stages. The use of correlation coefficients of CAR-characteristics and measures of child language progress often leads to the occurrence of spurious correlations which are difficult to explain. According to Gleitman et al. (1984) this has been the case in the

Furrow study. In a replication of the Furrow study they found a smaller number of significant correlations between CAR characteristics and language learning. Complex sentences in CAR were significantly positively correlated to the child's learning progress. On logical grounds, Gleitman et al. concluded that complex sentences are more informative to children acquiring language than simple sentences, because they contain more information and display a larger variety of lexical relations. In addition they found also some evidence that young children tend to be biased towards a stereotyped ordering, and, more specifically towards declaratives. According to them, this could explain why the child preserves this ordering for several other sentence types, in the way they are reordered at the surface of the CAR data. Gleitman et al. reported also that there were marginally significant effects of maternal expansions and effects of maternal sentence complexity on the child's growth in using auxiliary verbs. The same results were found in a similar study concerning the influence of CAR data on language learning by Barnes, Gutfreund, Satterly and Wells (1983).

They found some additional results in children's language learning progress, based on explicitly stated measures for child progress:
- In general, extending and contingent utterances, defined as those that pick up and elaborate on or add to the meaning the child has just contributed, are strongly associated with progress in language acquisition
- The frequency of directives, rather than questions calling for comment, are associated with child language learning
- Polar interrogatives, a result of inversion of the subject and the auxiliary verb (e.g. 'Have you finished breakfast?') or of intonation patterns only (e.g. 'You finished breakfast?') were also associated with specific measures of children's progress.

Cross-linguistic and cross-cultural studies mentioned earlier, together with the serious methodological limitations of the correlation studies discussed above, lead to the conclusion that we can say little more than that the statistical correlations between certain characteristics of adult language input and the child's language development is a plausible fact. However, it would be wrong to interpret these empirical (statistical) correlations (as in Wells et al., 1983) as established causal relationships between CAR and the child's general progress in language development (Cook & Campbell, 1977). The conflicting results and methodological problems of language input studies make it very difficult to apply the results from these studies to the explanation of specific CAR characteristics in conversations with SLI children.

3.2.3. Topic and Theme Management

The term 'discourse topic' appears frequently in current research on language, but there is no agreement on a precise definition.

Keenan and Schieffelin (1976; p. 343) defined topic as a :

"proposition (or set of propositions) expressing a concern

(or set of concerns) the speaker is addressing" .

Halliday and Hasan (1976) pointed out that the connections created by shared propositional content play an important role in organising discourse elements into a unified text or conversation. Hurtig (1977) stated that links between successive utterances in discourse are mostly based on shared propositional content and the use of identifiable linguistic references.

Bates and MacWhinney (1979) observed that the selection of a topic depends on the shared information between speakers as well as on previous discourse notions.

Keenan and Schieffelin (1976) were more specific in their formulation of what is to be considered a topic. They noted that in order to establish a topic in discourse, a speaker must meet some prerequisites including securing the listener's attention, speaking clearly, the use of identifiable referents and semantic relations between them. Keenan and Schieffelin focused their analysis of prerequisites on subsequent utterances. In the utterances following a topic introduction, the topic may or may not be continued. They specify which topic criteria are used for continuous and discontinuous discourse. Although researchers from several disciplines have studied the nature of topic manipulation in adult discourse, the development of topic manipulation patterns in child interactions has only recently been examined. The available research points at several differences between topic manipulation in adult and child language.

According to Keenan and Schieffelin (1976) young NLA children (from age 1;4 to 2;10), in interaction with their mothers, do not introduce a new topic adequately for reasons of limited attention span or distractability (e.g. the noise of a motorcycle passing by), or failure to grasp the point of the preceding utterance (e.g. misunderstandings and ambiguity). Children may also have some difficulty in judging the relevance of one utterance to another.

This could be explained as having difficulties with the comprehension of the context and with the use of the correct deictic pronouns and with the employment of other referential skills.

Keenan and Klein (1975) set up a longitudinal study on the linguistic interactions of a dyad of twin boys between the ages of 2;9 to 3;9 and they observed that these children used repetitions of each other's utterances (or parts of utterances) to provide some coherence in their interaction.

Bloom et al. (1976) studied adult-child interactions involving four children as they progressed from Brown's Stage I to Stage V (Brown, 1973; see also section 3.2.1.). They examined child utterances adjacent to or following adult utterances. They found that children produced utterances which shared the topic of the adult utterance or added new but related information. The information while maintaining the topic increased from Stage I to Stage V. Contingent speech increased over time; in particular linguistically contingent speech (speech that expanded the verb relation of the prior adult utterance by adding or replacing constituents within a clause) showed the largest developmental increase. Linguistically contingent speech occurred more often after questions than after non-questions.

Fine (1978) examined pupils from 5;0 to 9;0 involved in conversations with peers and with their classroom teacher. In this study it was found that the younger children link their own language to that of the teacher more often than the older children do.

Ervin-Tripp (1979) examined conversational samples of discourse from speakers in various situations and found that 2-year-olds were capable of maintaining a topic when replying to adjacency pairs such as choice questions, commands and offers.

She also found that children's ability to obtain a speaking turn in discourse was in part dependent on their ability to contribute relevant information, thus maintaining the topic.

Brinton and Fujiki (1984) investigated the manipulation of discourse topics in spontaneous conversation in three age groups.

Their subjects consisted of six dyads of peers at each of the following age levels: 5;0-5;11, 9;0-9;11 and adults. There were remarkable differences in topic manipulation between the age groups. The five and nine year old speakers re-introduced significantly more topics than the adults. Adults centered their conversations around fewer topics, whereas children moved quickly from topic to topic. Children seemed to be more likely to talk about concrete objects they happened to be attending to at the time.

The data further suggest a growing realisation of, and adherence to, relevant requirements in conversation. The ability to use topics to provide coherence in discourse appeared to increase with age. The topics maintained by adults consisted of more utterances than topic maintenance by children. Younger children used significantly more repetitions in order to maintain a topic. According to Keenan and Klein (1975), this reliance on repetitions appeared to serve two different goals :

- to seek or provide acknowledgement of a previous utterance
- to use repetition to provide some continuity or coherence in conversation.

Foster (1986) studied the ability of children at age 0;1 to 2;6 to initiate and maintain topics of conversation. An analysis of both verbal and nonverbal acts suggests that at the beginning of the development children simply attract more attention to themselves as topics of conversation.

In adult-child discourse children's utterances are interpreted by the caretaker as meaningful and connected with preceding events. In maintaining topics children are initially capable to extend sequences within the context of routines, but by age 2;6 they are able to maintain coherent topics independently.

However, the lack of sufficient research on normal (and deviant) developmental aspects of topic manipulation in discourse makes identification and interpretation of related disorders difficult (Brinton & Fujiki, 1984).

There has been considerable interest in and speculation about the possibility of specific topic manipulation in SLI children. Investigations have shown that the ability to initiate and maintain conversational discourse (in terms of topic and theme management) is established within the second year of life (Bloom et al.,1976; Keenan & Schieffelin, 1976; Foster, 1986; Chapman, 1981; Wanska & Bedrosian, 1985). Through interaction with the caregiver, a child learns to follow the organisational and sequential structure imposed by the adult and learns to exert some control over that structure.

Only a few studies have focused on topic management in studies with SLI and NLA children and their mothers (Fey & Leonard, 1983; Johnston,1986). The studies reported indicate that, compared with NLA children, SLI children have general constraints in the use of language and particularly in the use of cohesion devices as compared with NLA children (Liles, 1985).

From their empirical study aiming at testing a descriptive taxonomy of pragmatic skills, Prutting and Kirchner (1987) derived that SLI children (compared with other disordered groups and NLA children) have severe problems with cohesiveness,* the ability to revise conversation when a communication breakdown (e.g. misunderstandings, unfinished utterances, interruptions, acting incorrectly) occurred and the ability to ask for a repair (e.g. repetition or paraphrase of previous act, expansions, clarification requests) when a misunderstanding or ambiguity has occurred. However, the information gathered from the procedure and the protocol does not specify the kind of erroneous cohesion devices used.

According to their data, topic maintenance contains a larger number of utterances and turns when the topic is child-initiated and the conversational focus is an object in the direct environment. Conti-Ramsden (1983, 1988, 1989) reported differences in the use of cohesion illocutions (e.g. contingent responses, attention getting, clarification requests) between mothers of SLI and NLA children (between two and three years old). Since NLA children initiated dialogue

*Defined according to Halliday and Hasan (1976), see also section 2.3.2.1.

more often than SLI children, mothers differed in the number of utterances that provided information requested by choice questions (e.g. Child asks: "Do elephants eat spaghetti?" and his Mother answers: "No") and acknowledgements. The mothers of NLA children had more chances to respond to what their children said, to answer their children's questions and to acknowledge their children's comments. Mothers of SLI children had such opportunities less often. It appears that mothers of SLI children and NLA children differed mostly in the use of cohesion illocutions which depend on the child acting as an initiator.

3.2.4. *Communication Acts*

Speech is only one mode of communication, although it is a highly structured one. In communicative interaction it is mostly accompanied by or even (partly) replaced by nonverbal communication. In order to incorporate nonverbal as well as verbal acts in the functional description of discourse, the term 'speech act' is replaced by 'communication act' in this study (see also section 4.3.5.).

It seems reasonable to suppose that SLI children show no difficulties in their performance of communication acts despite significant limitations in their morpho-syntactic abilities. There is however not sufficient support for this presupposition (Morehead & Ingram, 1973). Most studies report deficient use in the range of communication acts in addition to problems in formal linguistic skills.

In their review Fey and Leonard (1983) summarised that SLI children reflect deficiencies in their ability to produce requests for action and information. In addition to that they mention that additional problems in the acquisition and use of anaphoric pronouns for encoding previously established referents could be seen as a cause for limitations in interpreting and performing cohesive and functional meaning relations in discourse. Keenan and Schieffelin (1976) reported a similar conclusion.

In a study conducted by Schwabe, Olswang and Kriegsman (1986) SLI children were restricted in their use of requests for information as a functional means to obtain new information. They conclude that these children lack the ability to inquire about different aspects of the environment, which limits their flexibility and efficiency in learning about the world and in developing language.

In a comparative study of the use of clarification requests by NLA and SLI children (mean chronological ages of 5;5 years) Hargrove, Straka and Medders (1988) found that SLI children had the capacity to provide feedback in the form of clarification requests when faced with failures in understanding correctly verbal and nonverbal acts of their mothers. SLI children tend to be less specific in their questioning than NLA children when requesting a clarification.

In the study of Brinton and Fujiki (1982) SLI children frequently ignored or responded inappropriately to adult requests. The responses given were occasionally contrary to or totally unrelated to the expected acts. These data give evidence of the fact that SLI children seem to have difficulties in grasping essential information from language input data and contextual events; they do not have full access to a range of linguistic tools needed for requesting essential information in order to create appropriate and coherent responses.

3.2.5. Communication Breakdown and Repairs
When individuals do not properly regulate or modify messages, a breakdown in communication often occurs. Communication breakdowns occur for many reasons (Roth & Spekman, 1984; Blau, 1986):
- unintelligible articulation
- improper loudness
- incomplete (unfinished) or ambiguous utterances
- overlaps and interruptions
- complexity
- inappropriate utterances and (more specifically) responses
- absence of cooperative partners.

When a breakdown in communication occurs, the listener often requests clarification. These requests have one of several realisations, including requests for confirmation, neutral requests, and requests for specific constituent repetition (Brinton et al., 1986; Prather et al., 1989). The speaker whose message is not understood may attempt to repair the breakdown. These repairs also have one of several forms, such as repetitions, revisions (e.g. correction with elaboration, substitution, reduction with substitution), neutral responses (e.g. confirmation, negation), additional references and questions. The literature concerning conversational repairs in communicative interaction of SLI children and their caretakers suggests that the problems these children have with understanding and creating coherent discourse will lead to an increase in communication breakdowns, such as incorrect responses of the children, repetitions and clarification requests of the caregiver (Gallagher, 1977; Gallagher & Darnton, 1978; Roth & Spekman, 1984; Brinton et al., 1986; Prather et al., 1989).

Prather et al. (1989) found a higher frequency of adult clarification requests directed to SLI children than to NLA children (at age 4;0 to 5;6). The SLI and NLA children were matched on age, cognition and verbal competence. According to these authors, adults try to repair inappropriate responses of the children and facilitate the use of complex repairs of the children.

The clarification requests then function as CAR. A relation of clarification requests to conversational repairs of communication breakdowns seems plausible (Gallagher, 1977; McTear, 1985; Brinton et al., 1986). Prather et al. (1989) found no significant differences in the type of repairs used by SLI and NLA children. SLI and NLA children most frequently used 'revisions' as repairs. This finding contrasts with the study of Gallagher and Darnton (1978), which indicates that SLI and NLA children differ in the number and types of repairs used. The study of Gallagher and Darnton and the study of Prather et al. differ with respect to the subject selection process, which could have affected the results. Gallagher and Darnton analysed the responses to requests for clarification of twelve language impaired children. The results were compared with the data from an earlier study of NLA children (Gallagher, 1977), in which it was found that older children use more complex responses and revisions (corrections and elaborations, reductions with substitutions). There seems to be a progression of revision behaviours with age. The language delayed children, although sensitive to revising communication breakdowns, used less sophisticated devices. The responses of the language delayed children were less elaborative than would be predicted by their acquired level of grammatical ability. Two main conclusions were drawn from this study:

- the language impaired children were delayed in language performance compared with the NLA children
- the language impaired children also showed qualitative differences with the NLA children in the use of conversational skills. The language impaired children failed in utilising their acquired grammatical knowledge for more sophisticated repairs.

Little more has been published about communication breakdowns and conversational repairs in conversations of SLI children and their caregivers. In general most studies suggest that SLI children will show a lack of awareness in the need to repair conversational breakdowns and will have difficulties in using repair strategies successfully (McTear, 1985).

3.3. Summary

In this chapter an overview was given of the research in discourse coherence conducted in the area of caregiver-child communicative interaction. Five main areas of research were discussed:

1. *Turn-taking*
   SLI children are willing to participate in communicative interaction with their caregivers, but they do not have full access to the required linguistic skills. The responsive behaviour of the SLI children mostly relies on the preceding acts of the caregivers (e.g. the use of ellipsis, repetitions, backchannels). The literature survey of 'turn-taking' will lead to the specification of six hypotheses (section 4.5.1.).

2. *Child Adjusted Register (CAR)*
   In general caregivers of SLI children seem to adjust their language input similar to the caregivers of NLA children. Most of these similarities concerned the MLU and sentence complexity of the caregivers' verbal language input. However there are also some differences in the CAR of caregivers communicating with their SLI and NLA child, mainly concerning the number of acts used per turn and the number of conversational repairs from the caregivers (in order to preserve discourse coherence). The literature survey of 'Child Adjusted Register (CAR)' will lead to the specification of three hypotheses (section 4.5.2.).

3. *Topic and Theme Management*
   SLI children seem to have specific difficulties with initiating and maintaining conversational themes. SLI children also showed difficulties in inquiring required information for sustained discourse from the caregivers. Caregivers of SLI children have fewer opportunities to respond to what their children say or do. The literature survey of 'Topic and Theme Management' will lead to the specification of five hypotheses (section 4.5.3.).

4. *Communication Functions*
   SLI children have problems in grasping essential information from language input data, the non linguistic situation and the contexts. For this reason it is expected that SLI children will show a reduced number of information requests and controls for regulating the discourse activities when compared with NLA children at the same age. The literature survey of 'Communication Functions' will lead to the specification of two hypotheses (section 4.5.4.).

5. *Communication Breakdowns and Repairs*
   Conversations between caregivers and their SLI children result more often in communication breakdowns than conversations between caregivers and their NLA children, due to a lack of awareness in the need to repair communication breakdowns and the difficulties SLI children have in using repair strategies successfully. Compared with caregivers of NLA children, caregivers of SLI children spend more time in securing discourse coherence (e.g. in the use of revisions, elaborations, repetitions). The literature survey of 'Communication Breakdowns and Repairs' will lead to the specification of seven hypotheses (section 4.5.5.).

The chapters 2 and 3 gave an overview of the literature on which the present study is based. The next chapter offers definitions of the linguistic categories used to code discourse coherence and incoherence. Some of the categories and some of definitions given in chapter 4 are derived from the discourse coherence studies reviewed in chapter 2.

CHAPTER 4

Interaction analysis of communicative abilities (IACV)

In this chapter we will look at some of the central organisational and structural aspects in spoken discourse which are relevant to our description and analysis of discourse coherence in children's acquisition of conversational abilities. The basic notion in the presented IACV discourse framework is that the verbal and nonverbal act is seen as the most elementary unit in communicative interaction. Conversational participants regulate and manage discourse at the verbal and nonverbal act level. The description of spoken discourse should reflect the dynamic process of conversation, showing how conversation is a joint activity resulting from the communicative interactional contributions of the participants. The IACV categories used for the coding of discourse coherence and incoherence in adult-child conversation are described in this chapter. IACV itself is specifically designed for the purpose of studying interactive language behaviour (verbal and nonverbal) from the perspective of both conversational partners and considers various levels of description and analysis. The structural units of discourse in IACV (such as verbal and nonverbal act, turn and turn-exchange, subsequence, topic and theme) are defined in this chapter. Other, more general terms (e.g. language, communication, conversation, discourse, text, coherence) were discussed earlier in chapter 1. The chapter concludes with an outline of hypotheses based on the literature reviewed in chapter 3.

4.1. Adult-child discourse in IACV

As is apparent from the literature in chapter 3, children acquire their language and communication abilities through interaction with those around them. This interpersonal context gradually helps the child to develop referential knowledge of objects and generates insight into the use of conversational rules and conventions.

The child is equally concerned with the acquisition of the linguistic system through which these meanings are expressed. The adult language input facilitates the child's language learning. The child needs to get sufficient opportunities to encode his own needs and meanings in linguistic forms already learned and to receive feedback on their appropriateness to guide further learning. Studying the way in which adult and child structure their conversational exchange reveals information about the way in which language and communicative abilities are acquired.

The interactional and interpersonal perspective of the present study requires the definition of some key-notions, such as language, discourse and communication (defined in sections 1.6. to 1.9.), and structural properties of the discourse, such as verbal and nonverbal act, turn and turn-exchange, subsequence, topic and theme (to be discussed in this chapter). These structural properties are the basic notions of the IACV framework. The theoretical basis of our IACV analysis model is eclectic, deliberately avoiding the attempt to construct an underlying model for communication analysis based solely on one of the competing theoretical approaches and representative models (as described in section 1.3. and chapter 2).

4.2. Units of Description and Analysis in IACV

In discourse the (language) codes in which the information is framed are sequentially ordered. Temporal organisation of communicative events is an important feature of communicative interaction. One of the first problems to face in the construction of a useful framework for interaction analysis is deciding which structural units of description and analysis should be used in studying spontaneous conversations between children and their caregivers. Many of the earlier described theories and models (chapter 2) offer a variety of possibilities, such as sentences, discourse constituent units (dcu), syntactic clause, speaker turn, context space.

These candidates have some major shortcomings:
- they are not coextensive
- they mostly reflect a preoccupation with spoken discourse and
- they are defined from a view which puts the speaker, actor or writer at the centre of the process of communication.

For the present not one of these units is more relevant or correct for the description and analysis of adult-child discourse. In fact, the choice has to be based on the specific purposes of the study. The solution adopted in this study is to use a hierarchy of units which permits discourse description and analysis at several levels of discourse organisation. The units selected are used for description as well as for analysis at morpho-syntactic level (clause, phrase, morphology) and pragmatic level (such as initiations, responses, correctness of language use, communicative function, topic shift). The units, in descending order of scope, are: Situation, Context, Topic, Subsequence (Theme), Turn, Verbal and Nonverbal Act, Clause, Phrase, Words and Syllables. The way in which the communication situation is organised in the IACV project is presented in figure 4.1.

FIGURE 4.1. Structural Organisation of Discourse in IACV

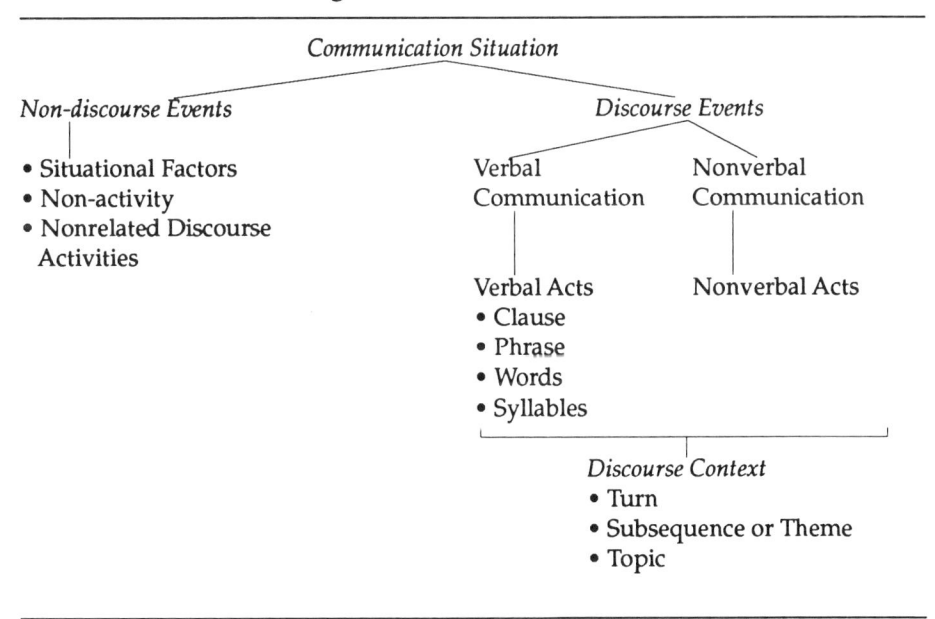

Communication Situation

Non-discourse Events
- Situational Factors
- Non-activity
- Nonrelated Discourse Activities

Discourse Events

Verbal Communication

Nonverbal Communication

Verbal Acts
- Clause
- Phrase
- Words
- Syllables

Nonverbal Acts

Discourse Context
- Turn
- Subsequence or Theme
- Topic

4.2.1. Situation and Context

Two major problems of the Generative and Descriptive Models for the study of discourse coherence, discussed in chapter 2, concerned the variability of the communication situation and the reactions of conversational partners. Any adequate theory of discourse coherence takes into account the role of the situation. Besides the effect of situational factors on the performance and interpretation of verbal and nonverbal acts, communication partners also rely on shared knowledge of the (language) code used, the topic of conversation and general knowledge of world. This knowledge generates conversational expectations about adequate and appropriate responsive behaviour. The term 'situation' refers to language independent information that influences the communicative setting. The term 'context' refers to the whole stretch of language dependent information framed in the preceding text.

Judging the appropriateness of acts is only possible within a known 'context' and with the help of situational knowledge. The appropriateness of a verbal or nonverbal act is context dependent and is achieved collaboratively by the conversational partners on each occasion. That is, what a speaker deems to be appropriate has to be endorsed (explicitly or implicitly) by the hearer or perceiver, or alternatively rejected (McTear, 1985, citing MacLure).

The situational and contextual information needs to be represented in the transcripts, because it functions as background information for the analyst when coding the functions of sequences of verbal and nonverbal acts.

4.2.2. Verbal and Nonverbal Acts

If any of the above-mentioned units is more basic than the others it is bound to be the verbal and nonverbal act*, as both are treated as minimal carriers of information. Discourse is locally managed at act level and guided by the interpretative processes of the communication partners. Thus, the structure of conversation is primarily realised at act level, which influences turn-taking behaviour, functional and formal properties of acts and topical organisation. All these aspects organise the information flow in conversations on which the child builds his strategies for acquiring communicative abilities and skills.

The analyst assigns the formal and functional properties of acts during the IACV analysis of the discourse transcripts. Discourse is realised as (verbal and nonverbal) acts, which in fact are the smallest possible interactionally meaningful contributions to conversation.

---

* In those cases where the term 'act(s)' is used without the adjective 'verbal' or 'nonverbal', verbal as well as nonverbal acts are meant.

Verbal Acts or Utterances

An utterance is seen as the verbal realisation of an act (Van der Geest, 1977) and has a pragmatic function (resulting in a communication act) with lexical and grammatical properties. We therefore prefer to use the word 'verbal act'. There are many other definitions of 'utterance'*. According to our view a verbal act may consist of a single word, clause or phrase; it even may consist of other (subordinated or embedded) constructions; it may be grammatically incorrect or unfinished (Brown & Yule, 1983; Lyons, 1977). The definition of a 'verbal act' or 'utterance' is derived from Hunt's T-unit:

> "one main clause plus any subordinate clause or nonclausal structure that is attached to or embedded in it" (Hunt, 1970,4).

Conjunctions are seen as separate utterances (main clauses), except contractions. Nonstructured parts, e.g. 'look', 'yes', 'no', are seen as part of the main clause. The segmentation of the transcript into verbal acts is also influenced by intonation patterns (e.g. a decrease in intonation in affirmative acts indicates the termination of the utterance) or changes in patterns of accompanying nonverbal behaviour, such as gazing, head movement, gestures and leg position (Duncan & Fiske, 1977). Besides structural and intonational criteria the contours of verbal acts can be, incidentally, detected by so-called interval time (Bol & Kuiken, 1988,23). Interval time is defined as the time elapsed between two verbal acts (Van Balkom et al., 1989,3). Within IACV a description of morpho-syntactical properties of verbal acts is given. Verbal acts of the children are analysed at clause level, phrase level and word level, using the GRAMAT**-procedure developed by Bol and Kuiken (1989). The adults' verbal acts are described in more general terms (e.g. grammatical complexity, the kind and number of verbs, the use of grammatically incorrect utterances and Child Adjusted Register (CAR)).

In some of the generative approaches of discourse coherence (in section 2.3) utterances were defined as functional units. Hartveldt (1987) states that the recognition of an utterance depends on recognising it as a speech act. Since coherence is a functional property of discourse, it is the communication act rather than the syntactic construction of a verbal act that we have to look for. We also saw, especially in the Generative Local Model approach of discourse coherence (section 2.3.2) that coherence is also established within an utterance through the use of cohesion devices by lexico-syntactic signals. Coherence covers morpho-syntactic (in verbal acts) and pragmatic properties (in verbal

---

* The word 'sentence' is often used as a synonym for 'utterance' in analyses of written discourse. In the study of spoken discourse we prefer the use of 'utterance' or in this study 'verbal act'.
** GRAMAT = *Gram*maticale *A*nalyse van *T*aalontwikkelingsstoornissen ('Grammatical Analysis of Developmental Language Disorders')

ánd in nonverbal acts) and occurs within and between verbal and nonverbal acts of both conversational partners. So, the study of discourse coherence in IACV considers morpho-syntactic categories (e.g. ellipsis, suppletion of unfinished verbal acts; nonstructured responses (e.g. "yes", "no")) as well as pragmatic functions (e.g. elaborations of conversational themes, imitations, repairs, backchannels).

Nonverbal Act
It is important to stress the role of nonverbal behaviour. Nonverbal behaviour includes gesture (e.g. pointing, gaze, body orientation, movements and intentional acts (e.g. the giving and taking). In the IACV project a distinction is made between communicatively intended nonverbal behaviour and unintended nonverbal behaviour. In many cases it is difficult to make this distinction. Consider, for example, gestures (e.g. pointing, giving, reaching) and eye movements which accompany speech but which occur largely below the level of the interpreter's consciousness and often are considered non-intentional. For example, gestures and (establishing) eye contact are important nonverbal means of regulating a smooth transition of turns in face-to-face interaction (Duncan & Fiske, 1977; McTear, 1985).

This important regulating and often initiating character of nonverbal behaviour in spoken discourse is a motivation for transcribing and analysing nonverbal behaviour in IACV.

Compared with verbal acts, the definition of a nonverbal act as an organising unit in discourse is difficult, because of the problems with making a clear distinction between communicatively intended and non-intended nonverbal behaviour.

There is also the problem of duration and segmentation of nonverbal acts. In contrast to defining the segmentation of verbal acts we can not rely extensively on other studies which present definitions of nonverbal acts in conversations. Most studies in adult-child discourse are convinced of the importance of nonverbal behaviour in conversations but seldom consider nonverbal acts in their recordings, transcripts and analyses (as in Wells, 1985 and McTear, 1985).

> "However, whilst recognising the importance of nonverbal behaviour for a full account of communication which is predominantly carried on through speech (...) makes no attempt to describe this behaviour, on the grounds that in the recordings the evidence necessary for such a description was (...) absent". (Wells, 1985, 61).

Before the IACV project started, several pilot-studies were conducted in order to define nonverbal acts. The following rules were set up:

- nonverbal acts are singular, main activities (e.g. 'TAKES doll', 'POINTS to car')
- if two or more nonverbal acts of one actor occur simultaneously and have the same communicative intention, they are seen as one nonverbal act (e.g. 'LOOKS and POINTS to dolls' house')
- an ongoing activity of one partner is seen as one nonverbal act and is transcribed only once, at the starting point. The follow-up of that activity is presented as context information (example 6)

(6)   *Example of transcribing ongoing activities*

<a. takes doll>

1. c. //kijkt naar activiteiten van a.
       '//watches mother's activities]'
2. a. //trekt pop kleertjes aan]
       <c. blijft kijken naar activiteiten a.>
       '//dresses doll]'
       '<c. keeps watching activities a.>'
3. c. NEE mamma]
       <c. wil niet dat a. poppetje aankleedt>
       'NO mummy]'
       '<c. doesn't agree with activities of a.>'
4. c. //datte
       <a. blijft doorgaan met poppetje aankleden>
       '//that'
       '<a. continues dressing doll>'
5. c. wijst in doos naar andere kleertjes]
       'points into box to other clothes]'

1 and 2 are counted once as nonverbal acts, further on they are represented in the context of 2 and 4.

a.	: *adult/caregiver*
c.	: *child*
//...]	: *start (//) and end (]) of overlap of consecutive acts*
<...>	: *context*
'...'	: *paraphrase in English*

- verbal and nonverbal act of one partner performed at the same time are transcribed in sequential order (see example 6, act 4 and 5)

61

- simultaneously occurring nonverbal acts of both partners or nonverbal and verbal acts of both partners are also presented in sequential order, because there is a change of actor or speaker (e.g. example 6, acts 1 and 2).

Overlaps of verbal and nonverbal acts of one or both partners are difficult to deal with, but need to be considered in any study of discourse coherence because both kinds of acts can have different pragmatic functions (e.g. acts 2 and 3 in example 6).

4.2.3. Exchanges and Moves

In fact, two kinds of structural units have been proposed for studying the organisation of conversation. Conversation analysts (section 2.1.1.) prefer the term 'adjacency pair', while the discourse analysts (section 2.1.2.) use the term 'exchange'. Adjacency pairs were discussed earlier (section 2.3.2.2.) and seemed to be unsatisfactory structural units for the purpose of this study. The 'exchange' is closely linked to the use of the term 'move'. A move can have several structural characteristics.

Some of the generative approaches of discourse coherence describe verbal and nonverbal acts as 'moves', because of their intentionally used, functional and organisational properties in discourse. A 'move' combines both aspects of content and organisation, whereas a verbal or nonverbal act is concerned mainly with content and only minimally (with respect to cohesion devices) with discourse structure and organisation.

An 'exchange' consists of at least one move by one speaker or actor initiating the exchange, and a second move by another speaker or actor who responds to the initiation.

The classification of moves in terms of initiations, responses and comments is central in the IACV pragmatic analysis. However, the characteristics of moves are not indicated as such in the IACV transcript. The transcript is complete, but as neutral as possible. The events which occur in conversations cannot all be described completely (Ochs, 1979; Ten Have, 1982). The difficulty with the term 'move' is that it reflects the integration of descriptive and interpretative aspects of discourse units, which may threaten the validity and the reliability in the coding of functional properties of discourse (see section 5.3.2.). This is the reason that we consider the assignment of characteristics of moves (e.g. initiations, responses, comments) as an integral part of the actual pragmatic IACV analysis, rather than of the transcription. The analyst has to judge the intention and function of each act from the perspective of its performer. This enables him to decide upon the appropriate structural and organisational meaning and the intention or purpose of the act based on the context of preceding and following verbal and nonverbal acts in the transcript.

4.2.4. Turns, Turn-exchanges and Backchannels
Within the IACV project the role of turn and turn-exchanges is of minimal relevance, because the turn is not considered a functional unit in our analysis of discourse. However, quantitative aspects of conversational turns are of interest to this study: the number of verbal and nonverbal acts per turn (called the 'turn-length') and per session, the total number of turns per session; the number of simultaneously occurring acts in the turns of both partners (e.g. parallel talk) per session. It is not easy to formulate a definition of 'turn'. There are serious demarcation problems between turns. Sacks et al. (1974) state that it would be a formal solution to define turn-exchange by 'speaker-change'.

Change of speaker is then change of turn. Sacks et al. (1974) specify rules and conventions for 'speaker-change'. They conclude that 'speaker-change' is locally organised (at verbal act level) and influenced by interactional behaviour of both partners.

Matarazzo and Wiens (1972) define a turn as :
> "The total duration of time it takes a speaker to emit all the words that he is contributing in that particular unit or exchange". (1972,6)

Duncan & Fiske (1977) also mention the hearer in their definition of conversational turn :
> "a speaker is a participant who claims the speaking turn. An auditor is a participant who does not claim the speaking turn at a given moment". (1977,177)

But speakers can take the turn without aiming to wresting the conversational floor by using so-called 'backchannels'. Backchannels are verbal or nonverbal acts which function as conversational feedback of the hearer or receiver to the speaker or the actor. Backchannels are important for the current speaker. He draws essential information from these feedback signals concerning the acceptance of his acts and consequently obtains information to revise them. Backchannels may include nonverbal signals (e.g. frowning, nodding, smiling), brief vocal insertions (e.g. hmm, yeah) and longer verbal acts (e.g. restatements, clarifications, questions, agreements).

The above mentioned definitions of 'turns' are set up from a viewpoint which puts the speaker in the centre of the process of communication. Within the IACV project a 'turn' is defined as a segment in discourse in which one speaker or actor makes his contribution, verbal or nonverbal, to the conversation. For practical reasons and for the implementation of computer-aided analysis in IACV this definition implies that the change of speaker or actor results in a change of turn. Thus, backchannels count also as separate turns. In IACV, the turn is not seen as a category of discourse function, but of discourse structure.

The turn may be of importance as one source of information on the level of interpersonal coherence (e.g. the differences in length of turns of the partners). Turn-taking is locally managed, because it operates with one speaker or actor at a time. Turn-exchange is also managed interactionally, because it depends on the collaborative activities of the participants for its smooth functioning (Sacks et al., 1974). There is some evidence that children, before they reach school-age (4;0), acquire some essential and complex aspects of the turn-exchange process in conversation (McTear, 1985).

The definition of 'turn' raised specific problems concerning the sequential organisation of simultaneously occurring turns:
- simultaneously occurring verbal and nonverbal acts of the conversational partners (or 'parallel acts') were counted as separate turns

(7)   *Example of parallel acts and turn-exchange*

<c. needs to go to the toilet>

1.a.   //loopt naar de tafel
<a. wil iets anders gaan pakken>
'//walks to table'
'<a. wants to change play-materials>'
2.c.   Mamma ik moet naar de W.C.]
'Mamma I have to go to the toilet]'

---

a.   : *adult/caregiver*
c.   : *child*
//...] : *start (//) and end (]) of overlap of consecutive acts*
<...> : *context*
'...' : *paraphrase in English*

- backchannels were counted as separate turns. But at the end of each session the total number of backchannels is substracted from the total number of turns, because backchannels do not function and are not intended as real changes of turn.

4.2.5. *Subsequences and Themes*

If discourse is locally managed at act level, this means that the communication partners can change the function and organisation of discourse with the occurrence of a next verbal or nonverbal act. The plans according to which conversational partners use a communication act as an instrument are too many to enumerate. The ways in which verbal and nonverbal acts lead to the actual linguistic performance are represented in the transcripts and the functional properties of these acts are coded separately during the pragmatic analysis.

As Wells (1985,61) states, there would probably be general agreement that the list of functional categories, or communication acts* should include at least the following: to control action, to seek and to give information, to express emotion and to establish and maintain social relations. The speaker's or actor's specific intention (plan) and the linguistic instrument with which this plan will be performed is usually the responsibility of the initiator of the exchange. If the other participant in conversation is willing to accept these actions, the discourse will develop smoothly and coherently. In case the other partner does not accept the performance of the speaker's or actor's plan, the discourse will make frequent changes of direction or fail to reach recognisable conclusions. A stretch or unit of discourse on which the partners mutually agree with respect to the purpose and current topic is called a 'subsequence' (according to Wells, 1985,62) and has a specific communicative, thematic function (e.g. to control action, to seek and to give information, to maintain social exchange) or 'theme'. The 'thematic content' of the 'subsequence' is established in the use of specific communication acts. So, a subsequence consists of an initiating verbal or nonverbal act and all the subsequent verbal and nonverbal acts depend on it until the next initiating act. Subsequences and their thematic content are delimited by consecutive initiations (see example 8, after section 4.2.6.).

4.2.6. *Topic of discourse*

Thematic continuity is fundamental to smooth efficient and coherent conversational exchanges (Tracy, 1985). 'Topic' is a continuously changing entity, modified at the level of verbal or nonverbal acts. In the IACV project 'topic' is seen as the concatenation of themes serving the same general idea. A 'topic' is seen as the main point of conversation. The thematic content of subsequences, the 'theme', is seen as episodes or concrete examples of the more general 'topic'.

---

* In order to incorporate nonverbal as well as verbal acts in the functional description of discourse the term 'speech act' is replaced by 'communication act' (see also section 3.2.4.).

(8)   *IACV Analysis Example, showing the patterning of discourse in verbal and nonverbal acts, turn-exchanges and themes*

<a. and c. are playing together>

TOPIC: PLAYING WITH DOLLS AND DOLLS' HOUSE
―――――――――――――――――――――――――――― START THEME 1
1.a. //kijk die heeft een paardestaart   i
     'look, she <the doll> has a pony-tail'
2.a. wijst op poppetje]   cm
     'points to doll]'
―――――――――――――――――――――――――――― START THEME 2
3.a. och: en hier een babytje   i
     'we have a little baby here'
4.a. pakt poppetje uit poppenhuis   cm
     'takes doll from the dolls' house'
―――――――――――――――――――――――――――― START THEME 3
5.c. OH: //ja kijk maak (ik) de deur dicht   i
     'OH: //yes. Look, I close the door'
6.c. maakt deurtje dicht]   cm
     'closes the door of the dolls' house]'
―――――――――――――――――――――――――――― START THEME 4
7.a. Oh: //en nou moeten ze naar bedje   i
     'Oh: //and they have to go to bed now'
8.a. staat op van vloer]   cm
     'rises from floor]'
9.a. //en nou moeten ze naar bedje brengen?   cm
     '//and they have to be put to bed now?'

―――――――――――――――――――――――――――――――――――――
i   : *initiation (see section 4.3.1.)*
cm   :*comment (see section 4.3.2.)*
a.   : *adult/caregiver*
c.   : *child*
//...] : *start (//) and end (]) of overlap of consecutive acts*
<...> : *context*
...'   : *paraphrase in English*
CAPITALS (e.g. 'OH') : *stressed word or wordpart*
―――――――――――――――――――――――――――――――――――――

4.3. Coding Discourse Coherence in IACV

The IACV coding scheme includes a vast number of morpho-syntactic and pragmatic categories. In the next sections definitions are given of a set of IACV categories which have been used for the study of discourse coherence. These definitions are fairly exploratory and by no means watertight. This implies that the definitions of terms and categories used by the analysts have to be formulated and exemplified clearly. The analyst applies these definitions to recognise the status of verbal and nonverbal acts as they occur at any given point in interaction. As stated so often in the preceding sections and chapters, conversation is seen as a joint activity which is created on an act-by-act and turn-by-turn basis. The fact that the definitions given are not watertight reflects the indeterminate nature of adult-child discourse, for, in many cases, analysts can disagree as to whether a given act requires a further response or not (see also McTear, 1985, 38-39).

4.3.1. Initiations

Verbal and nonverbal acts which are coded as 'initiations', are prospective; that is, they set up expectations of the initiator about what type of responses or comments can be expected. Initiations serve as starting points for conversational exchanges and indicate the specific content and purpose of the subsequence. This functional entity is referred to as 'theme'. Besides the group of initiations which is easy to determine (e.g. questions, acts which change conversational themes) the following acts are also counted as initiations in IACV:

- backchannel-questions (e.g. "what?", "isn't it?", frowning)
- attention-getting and attention-directing (e.g. "Linda, look at this", pointing, touching the hearer, to give something to the hearer, initiating eye-contact, movements towards the hearer, greetings, screaming, whispering)
- corrections of previous acts (e.g. act 2 in example 9)

(9) *Example of correction*

<a. and c. play with tea-set>

1.c. datte koek
 'that is biscuit'
2.a. dat is geen koek maar een koe
 'that isn't a biscuit, it is a cow'

- acts which do not function as a clear response or comment on a previous initiation (e.g. all acts in the next example)

(10) Example of acts, not related to a previous initiation

    <a. and c. are playing separately>

    1.a. waar zijn de boeven?
          <a. speelt met boerderij>
          'where are the thiefs'
          '<a. plays with farm>'
    2.c. speelt met de dieren
          <achter de boerderij op de vloer>
          'plays with the animals'
          '<behind the farm on the floor>'
    3.a. Hé Linda
          <aandachtstrekker>
          'Hé Linda'
          '<attention-getting>'

---

a.	: *adult/caregiver*
c.	: *child*
<...>	: *context*
'...'	: *paraphrase in English*

There are three possible initiation-moves:
- initiations which introduce a new theme, including acts reacting to events outside the conversational context (e.g. act 2 in the next example).

(11) Example of reaction to events outside discourse context

    <a. and c. are playing with swimming pool>

    1.c. zet glijbaan op zwembad
          <stoot poppetje en boompje om>
          'puts slide on swimming pool'
          '<pushes down doll and tree>'
    2.a. Tim, wat doe je nu?
          'what are you doing now, Tim?'

- initiations which elaborate on a current theme; including back-channel-questions

(12)   *Example of initiation, elaboration theme*

<a. and c. playing with farm>

1.a.	//dat is een koe '//that is a cow'	i-new
2.a.	wijst naar dier op vloer] 'points to animal on floor]'	cm
3.c.	ja 'yes'	r
4.a.	of niet? 'isn't it?'	i-ela
5.c.	ja 'yes'	r

---

i-new   : initiation, introducing a new theme
i-ela   : initiation, elaborating on a current theme
r   : response
cm   : comment
a.   : adult/caregiver
c.   : child
//...]   : start (//) and end (]) of parallel acts
<...>   : context
'...'   : paraphrase in English

- initiations which reintroduce a theme after an earlier introduced (new) theme. A speaker or actor who uses a reinitiation of a theme tries to secure a satisfactory response. Reinitiations are also used to resume a topic after being interrupted by a situational event (e.g. the ringing of a phone, the noise of a motor-cycle passing by, someone knocking on the door).

Reinitiations are important developmentally as they give insight into the children's ability to pinpoint the possible reasons why their acts have failed to secure a response and the means they adopt to remedy this (McTear, 1985).

(13)   *Example of reinitation (reintroduction of theme)*

<a. and c. are playing with swimming-pool>

1.c.   laat poppetje in water duiken
       'makes doll dive into swimming pool'
2.a.   //kan jij ook duiken?
       '//you can dive too?'
3.c.   zet poppetje naast zwembad]
       'puts doll next to the swimming pool]'
4.a.   jij kunt toch ook duiken?
       'you can dive too, can't you?'
5.c.   laat poppetje zwemmen
       'makes doll swim'
6.a.   laat eens zien hoe je kunt duiken
       'show me how you can dive'

The acts 2, 3 and 5 are coded as elaborations of theme; the acts 4 and 6 are coded as reinititations.

a.	*: adult/caregiver*
c.	*: child*
<...>	*: context*
//...]	*: start (//) and end (]) of parallel acts*
'...'	*: paraphrase in English*

4.3.2.   Responses and Comments

Verbal and nonverbal acts which are coded as 'responses' are retrospective; they fulfil the predictions set up by a preceding initiation (see example 12, act 5). Verbal and nonverbal acts which are coded as 'comments' are neither prospective nor retrospective. They do not predict a further response, nor can they be seen as an elicited response to a preceding act. Comments describe preceding utterances, nonverbal acts or situational facts. They can serve as techniques to sustain communicative interaction (e.g. see example 8, acts 2, 4, 6 and 8; example 12, act 2)

Direct repetitions of initiations and simultaneously occurring acts with the same communicative intention are also coded as comments (see example 8; act. 9).

In Discourse Analysis a fourth type of move is proposed, both prospective and retrospective (McTear, 1985; Haft-Van Rees, 1989).

This move is coded R/I, a move which simultaneously functions as Response and as Initiation.

(14)   Example R/I (Response/Initiation) Move

1.c.	//wat is nou mamma? '//what is that mamma?'	I
2.c.	toont krokodil aan a.] <uit speel-boerderij> 'shows crocodile to a.]' '<from play-mobile animalfarm>'	
3.a.	wat zou dat zijn? <bang> 'what could that be?' '<afraid>'	R/I

a.	: *adult/caregiver*
c.	: *child*
//...]	: *start (//) and end (]) of parallel acts*
<...>	: *context*
'...'	: *paraphrase in English*

In this example the coding of functions of the move of the adult (3) could result in two communicative functions:
- a 'content question' (according to the syntactic structure of the verbal act of the caregiver; a Wh-question and
- a 'question with known answer*'(according to the communicative intention of the caregiver who actually knows the name of the animal (crocodile) shown by the child).

To deal with this problem of two possible functional codings, the caregiver's act (3) is characterised as an initiation (question with known answer) and the preceding child act in example 14 (act 1) is also characterised as an initiation (a 'content question').

Both the act of the child and the act of the adult are coded as initiations and not as an initiation-response sequence because the acts have different communicative intentions. The child's act is a request for specific information and the caregiver's act has a didactic function.

* (see Appendix C, for more specific definitions)

4.3.3. Ellipsis

Ellipsis or elliptical responses are answers to preceding questions without a main verb or a main clause.

(15a) Example of Ellipsis	(15b) Example of Ellipsis
1.a. vond je het leuk? 'did you like it?' 2.c. ja 'yes'	1.a. wat heb je gezien? 'what did you see?' 2.c. tijgers 'tigers'

Acts 2 are elliptical responses to acts 1

a.   : adult/caregiver
c.   : child
//...] : start (//) and end (]) of parallel acts
<...> : context
'...' : paraphrase in English

4.3.4. The Relation with Previous Acts

The pragmatical analysis categories in IACV apply to verbal and nonverbal acts of both partners. The information concerning the relation of an act to its previous acts and to the conversational partner is not represented in the acts and has to be derived from the linguistic context. Communication acts are related either to acts of the speaker/actor (referred to as 'self-related acts'), to acts of the conversational partner (referred to as 'partner-related') or are not related at all (e.g. situational events or the start of a new topic).

(16)   Example of partner-related act

<a. and c. are playing cards>

1.c. pakt kaart van vloer
     <verkeerde kaart>
     'takes card from floor'
     '<the wrong one>'
2.a. nee Saskia, dat is de verkeerde
     'no Saskia, that is the wrong one'

Act number 2 is 'partner-related'

(17)   *Example of self-related act*

   <a. and c. play at skittles>

   1.a.   gooit bal
          <mist>
          'throws ball'
          '<misses>'
   2.a.   dat was een slechte
          'that was a bad one'

   Act number 2 is 'self-related'

   a.   : *adult/caregiver*
   c.   : *child*
   <...>   : *context*
   '...'   : *paraphrase in English*

4.3.5. Imitations

Imitations consist of verbal or nonverbal acts of the child and caregiver that share the same conversational theme with the preceding act or an act within a range of three preceding acts and do not add (new) information. Imitations can either be related to acts of the speaker/actor (self-related) or to those of the conversational partners (partner-related). The group of imitations includes paraphrases or rephrasings of original verbal acts, expansions and repetitions.

(18)   *Example of imitation, repetition of partner's act*

   1. a. trek je trui uit Dave
          'take off your sweater, Dave'
   2. c. trui uit
          'sweater off'

(19)   *Example of imitation, rephrasing verbal act partner*

   <playing with dolls' house>

   1. c. dat is nolle
          <'nolle'=idiosyncratisch voor klok, via a.>
          'that is nolle'
          '<nolle=idiosyncrasy for clock, via a.>

2. a. ja dat is een klok
'yes, that's a clock'

(20)　　Example of imitation, expansion verbal act partner

<playing with dolls>

1. c. pop mam
'doll mum'
2. a. ja dat is jouw pop
'yes that's your doll'

The number 2 acts are coded as imitations

---

a.	: adult/caregiver
c.	: child
<...>	: context
'...'	: paraphrase in English

4.3.6. Lexical Coreference

The verbal acts of one conversational partner in which alternative words or circumscriptions are used to refer to earlier introduced words or names are called 'lexical coreferences' (Wells, 1985; Halliday & Hasan, 1977, 174)

(21)　　Example of Lexical Coreference

<a. and c. are playing with dolls>

1.a. //wat is dat?
'//what is that?'
2.a. wijst op hond]
<dierenpop bij poppenhuis>
'points to dog]
'<animal-doll near dolls' house>'
3.a. dat is Tessa
<naam van hun eigen hond>
'that is Tessa'
'<the name of their own dog>'
4.a. waar is ze naar toe?
'where's she gone?'

In act 4 'she' is coded as lexical coreference

4.3.7. Communicative Functions

Subsequences are classified according to the dominant purpose they take in the discourse (Wells, 1975, 1985):

- *Control*: the control of the present or future behaviour of one or both participants

(22)   Example of Control Subsequence

<c. wants to take new toys>

1.a. Linda eerst de autootjes opruimen
'Linda, clear away the cars first'
2.c. ja
'yes'

- *Expressive*: the expression of spontaneous feelings

(23)   Example of Expressive Subsequence

<a. gave a candy to c.>

c.   geeft kusje aan a.
'kisses a.'
a.   oh dank je wel
'oh thank you'
a.   jij bent lief
'you are nice'

---

a.	: *adult/caregiver*
c.	: *child*
//...]	: *start (//) and end (]) of parallel acts*
<...>	: *context*
'...'	: *paraphrase in English*

- *Representationals*: the requesting and giving of information

(24)   Example of Representational Subsequence

    c.   wat is dat?
           'what is that?'
    a.   dat is een puzzel
           'that is a puzzle'

---

    a.   : *adult/caregiver*
    c.   : *child*
    <...> : *context*
    '...'   : *paraphrase in English*

- *Social*: the establishment and maintenance of social relationships (e.g. greetings)
- *Tutorial*: interaction where one of the acts of the participants serves a didactic purpose.

(25)   Example of Tutorial Subsequence

    <playing with colour-pencils>

    a.   //welke kleur is dat?
           //'which colour is that'
    a.   toont potlood]
           'shows pencil]'
    c.   blauw
           'blue'
    a.   nee dat is niet blauw
           'no that isn't blue'
    a.   dat is groen
           'that's green'

- *Procedural*: management of the channel of communication and rectification of breakdowns due to mishearing or misunderstanding.

(26)   *Example of Procedural Subsequence*

<playing with cars>

a.   welke auto bedoel je?
     'which car do you mean?'
a.   pakt auto uit doos
     'takes car from box'
a.   //bedoel je deze?
     '//do you mean this one?'
a.   toont auto aan c.]
     shows car to c.]

a.	: *adult/caregiver*
c.	: *child*
<...>	: *context*
//...]	: *start (//) and end (]) of parallel acts*
'...'	: *paraphrase in English*

Each subsequence consists of acts serving a specific communication function (as specified above). These acts are classified in terms of communication functions which are specific for the thematic purpose of the subsequence (e.g. controls, tutorials). Definitions and examples of the possible communication functions for subsequences are given in Appendix C.

4.4.   Coding Discourse Incoherence in IACV

In this section a description is given of the IACV categories used for the coding of discourse incoherence, communication breakdowns and repairs. A study of discourse coherence is also a study of discourse incoherence. The purpose of this study does not lead to a description of grammatical categories, structural nonfluencies (e.g. self-corrections, word repetitions, false starts) or phonetic nonfluencies (e.g. unintelligible speech), although information about these aspects is available in the data resulting from the IACV analyses. When a conversation breaks down, the information flow from one partner to the other is disrupted. There are several ways in which communication can break down (e.g. interruptions, contrasting intentions) and there also are various types of repairs (such as clarification requests, reinitations, corrections).

This study is primarily interested in conflicting conversational interests and intentions of the partners, and in the ways in which these conflicts are repaired. The following sections briefly discuss some of the pragmatic aspects of discourse incoherence and repairs which are used in this study.

4.4.1. Unfinished Verbal Acts

Unfinished utterances often occur as a result of interruptions from the conversation partner or indicate that the speaker reconsiders the intended utterance. In the latter case the speaker often corrects or reformulates the previously unfinished utterance.

(27a) *Example of Unfinished Verbal Act/Interruption*

   <a. and c. are playing with telephone>

   1.c. //ikke foon
       'I phone'
   2.c. wijst naar 'Fisher-Price'-telefoon]
       'points to 'Fisher-Price'-phone]'
   3.a. dat is niet jouw //te-
       'that is not your //pho-'
   4.c. is] foon van mij
       'is] my phone'

Act 3 is unfinished because of interruption by child

a.	: *adult/caregiver*
c.	: *child*
//...]	: *start (//) and end (]) of parallel acts*
<...>	: *context*
'...'	: *paraphrase in English*

(27b) *Example of Unfinished Verbal Act/Elicitation*

   <a. and c. are playing with animalfarm>

   1.a //pakt beest uit doos
       <doos naast a. op vloer>
       '//takes animal out of box'
       '<box besides a. on floor>'
   2.a. //dit is een -]
       '//this is a -]'

3.c. kijkt naar beest]
<getoond door a.>
'looks at animal]'
'<shown by a.>'

Act 2 is unfinished because of elicitation of the adult

4.4.2. Simultaneously occurring Acts or Overlaps

Overlaps in conversations occur where more than one speaker talks at the same time (called 'parallel talk'), as in interruptions, or where more than one activity takes place. The acts involved in overlaps can have the same communicative intentions (e.g. the first two acts of the adult in example 25 and 27a) or have different communicative intentions (e.g. the acts 2 and 3 in example 13). Overlaps of verbal and nonverbal acts either are related to the speaker, actor (called 'self-related') or the communication partner (called 'partner-related').

(28) *Example Self-related verbal-nonverbal overlaps*

<playing with child 'medical kit' (=toy)>

1.c. //wat is dat mamma?
'//what is that, mamma?'
2.c. wijst naar stethoscoop]
<in de hand van moeder>
'points to stethoscope]'
<in mother's hand>
3.a. //de dokter luistert hiermee naar je hartje
'//the doctor uses this to listen to your heart'
4.a. pakt telefoontoestel van de tafel]
'takes phone from the table]'

Acts 1 and 2: The same communicative intention, self- related.
Acts 3 and 4: Different communicative intentions, also self-related.

---

a.	: *adult/caregiver*
c.	: *child*
//...]	: *start (//) and end (]) of parallel acts*
<...>	: *context*
'...'	: *paraphrase in English*

4.4.3. Incorrect Language Use

It is necessary to distinguish between the judgments of appropriateness of acts coded by the analysts post hoc and the remarks concerning appropriateness made by the conversational partners. Conversational partners recognise that they are being asked for information and are the recipients of a specific breakdown in the conversation. They usually say something relevant to what their partner has just said or done, and usually expect their partner to say or do something in return. The analyst has to follow the conversational partners in their activities in order to judge the overall appropriateness of the acts they used.

Within IACV correctness or appropriateness is coded for the morphosyntactic structure in verbal acts and the way in which relations with preceding acts are established in verbal and nonverbal acts (called 'use incorrect'). Especially 'use incorrect acts', such as faulty responses (e.g. the absence of expected responses) and faulty initiations (e.g. disregarding partner-related acts of the partner through continuing own activities) are of interest.

(29a) Example of Use Incorrect Acts (Faulty Initiations)

<a. and c. play cards>

1.a. toont kaartje aan c.]
 'shows card to c.]'
2.a. wat is dit?
 <kijkt naar c./ toont kaartje>
 'what is this?'
 '<watches c./shows card>'
3.c. waar is Judith?
 <zusje van c., zit in aangrenzende kamer>
 'where's Judith?'
 '<younger sister of c., in adjacent room>'

Act 3 is coded as 'use incorrect' (a faulty initiation)

a.   : *adult/caregiver*
c.   : *child*
//...] : *start (//) and end (]) of parallel acts*
<...> : *context*
'...' : *paraphrase in English*

(29b)   *Example of Use Incorrect Acts (Faulty Response)*

<a. and c. playing with farm>

1.c. //koe slapen nou?
'the cow is sleeping now?'
2.c. wijst op beest in boerderij']
'points to animal in farm']
3.a. nee dat is een paard
<kijkt naar c.>
'no, that is a horse'
'<watches c.>'

Act 3 is coded as 'use incorrect' (a faulty response)

a.   : *adult/caregiver*
c.   : *child*
//...] : *start (//) and end (]) of parallel acts*
<...>   : *context*
'...'   : *paraphrase in English*

4.4.4. Consecutive initiations

Consecutive initiations often indicate a communication breakdown:
- the speakers/actors of the initiations are involved in different, non-related activities and want to direct the partner's attention to his own current activity
- an initation of one partner is followed by a clarification request of the other partner (see example 26)
- an initiation of one partner is corrected by the other partner (see example 9)
- a speaker/actor reinitiates his preceding initiation
- an initiation is not followed by an expected response
- the speaker/actor has to repeat or adjust his initiation in order to make a response more likely (see example 13).

4.5. Hypotheses

The review of the literature in chapter 3 leads to a set of research hypotheses for the present study. The hypotheses are stated as 'alternative hypotheses' and not as 'null hypotheses', in order to indicate the direct relation to the results of the literature. The hypotheses reflect the suppositions (stated in the literature) concerning the differences in conversational participation and, specifically in the use of devices to create discourse coherence in conversations of caregivers and their NLA and SLI children. A survey of previous studies related to the presumptions stated in the hypotheses of the current study were discussed in chapter 3 (Turn-taking, section 3.2.1.; Social Context and Child Adjusted Register (CAR), section 3.2.2.; Topic and Theme Management, section 3.2.3.; Communication Acts, section 3.2.4.; Communication Breakdowns and Repairs, section 3.2.5.). We will not repeat all literature sources mentioned. The hypotheses raised in the next sections are sufficiently accounted for in chapter 3.

4.5.1. Turn-taking

Most of the literature on turn-taking related to SLI children indicates that SLI children want to participate in communicative interaction, but that they do not have full access to required linguistic skills. Their responses mostly rely on preceding acts of the caregivers (e.g. the use of elliptical answers, repetitions) and the use of backchannels (Roth & Spekman, 1984; Fey & Leonard, 1983; Demaio, 1982; Lasky & Klopp, 1982; Fey et al., 1981; Watson, 1977).

The results of the studies on turn-taking related to SLI children lead to a set of hypotheses for the present study.
Compared with NLA children of the same age and their caregivers :
1. SLI children use more backchannelling and ellipsis
2. caregivers of the SLI children take longer turns
3. caretakers of SLI children initiate or reinitiate themes more often
4. caretakers of SLI children use more self-repetitions
5. SLI children use more imitations of preceding adult acts
6. initiative behaviour of SLI children is mostly nonverbal.

The IACV categories used to find (counter-) evidence for these hypotheses are presented in the next table. Definitions of IACV categories mentioned are given in the previous sections. In this table the hypotheses have been indicated in the left column; the IACV categories associated with these hypotheses are described in the right column. Similar tables are used for the other four research areas.

TABLE 4.1 Turn-taking and associated IACV categories

Hypotheses	IACV categories
1. Backchannelling	Total number of acts of the child coded as backchannels
Ellipsis	Total number of the child's elliptical answers
2. Mean length of turn	Total number of acts of the adult divided by the total number of turn-exchanges of the adult
3. Adult (re)initiations	Total number of adult initiations and (re)initiations.
4. Self-related imitations adult/caregivers	Total number of adult/caretaker imitations, self-related
5. Partner-related imitations of child	Total number of partner-related imitations of the child
6. Nonverbal initiations	Total number of nonverbal initations of the child

4.5.2. Child Adjusted Register (CAR)

Although there is a growing general agreement among child language researchers about existing similarities in the ways in which adults adjust their language input to NLA and SLI children (e.g. grammatical and phonological simplifications, repetitions, elaborations) (Fey et., 1981; McDonald & Pien, 1982; Leonard, 1986), there also are some differences in Child Adjusted Register (CAR) used with NLA and SLI children (e.g. the use of repetitions of own utterances by the caregivers) (Lasky & Klopp, 1982; Schodorf & Edwards, 1983; Conti-Ramsden, 1983, 1988). The results of these studies on CAR relating to NLA and SLI children lead to a set of hypotheses for the present study.

Compared with NLA children of the same age, SLI children receive:
1. a language input characterised by a smaller MLU
2. a language input consisting of more (adult) self-imitations
3. a language input consisting of more (adult) requests for information.

The IACV categories needed to find (counter-) evidence for these hypotheses are presented in the next table (table 4.2). In this table the hypotheses have been indicated in the left column; the IACV categories associated with these hypotheses are described in the right column.

TABLE 4.2. Child Adjusted Register (CAR) and IACV categories

Hypotheses	IACV categories
1. Smaller MLU	MLU of the adult and of the child
2. Self imitations adult	Total number of self-imitations of acts by the adult
3. Adult's requests for information	Total number of requests for information by the adult.

4.5.3. *Topic and Theme Management*

The maintenance of topic is important in the construction of coherent discourse. Children seem to learn this in conversations with their caregivers mainly through the adults' use of cohesion illocutions, e.g. the getting of attention, requests for information, lexical coreference (Keenan & Klein, 1975; Bloom et al., 1976; Ervin-Tripp, 1979; Brinton & Fujiki, 1984; Foster, 1986). As already indicated in the hypotheses concerning 'Turn-taking' and 'Child Adjusted Register (CAR)', SLI children seem to be limited in the use of cohesive devices (e.g. the use of initiations to elaborate a conversational theme).

These problems with cohesion will lead to difficulties in maintaining conversational themes and topics and inquiring information of the environment (Brinton & Fujiki, 1984; Liles, 1985; Johnston, 1986). In addition it is also suggested that SLI children will initiate discourse less often than NLA children. As a consequence, caregivers of SLI children have fewer opportunities to respond to what their children say or do (Conti-Ramsden, 1983, 1988, 1989).

The results of these studies on topic and theme management relating to NLA and SLI children lead to a set of hypotheses for the present study. Compared with NLA children of the same age, the communicative interaction between SLI children and their caregivers is characterised by :
1. more child initiations of 'introduction new themes'
2. shorter themes in general
3. shorter themes initiated by caregivers of SLI children
4. more imitations of adult acts by the child
5. more imitations of self-related acts by the adult/caregiver.

The IACV categories needed to find (counter-) evidence for these hypotheses are presented in the next table (Table 4.3). In this table the hypotheses have been indicated in the left column; the IACV categories associated with these hypotheses are described in the right column.

TABLE 4.3 Topic and Theme Management and associated IACV categories

Hypotheses	IACV categories
1. Initiation new themes	Total number of initiation new themes by adult and child
2. Mean length theme	Total number of acts devided by total number of themes
3. Child initiated themes/ Adult initiated themes	Total number of acts devided by number of themes of child and adult
4. Imitations of adult acts by child	Total number of imitations of caregiver's acts used by the child
5. Imitations of self-acts by the adult	Total number of self-related imitations by the adult

4.5.4. Communication Functions

From several studies concerning the use of communication functions in adult-SLI child interaction, it appears that SLI children have problems in grasping essential information from language input data, the non-linguistic situation and the context (Schwabe et al., 1986; Hargrove et al., 1988). The results of studies on communication functions used by NLA and SLI children lead to a set of hypotheses for this study. Compared with NLA children of the same age, SLI children:
1. use fewer requests for information
2. receive a language input with more controls and tutorials.

The IACV categories needed to find (counter-) evidence for these hypotheses are presented in the next table (Table 4.4). In this table the hypotheses have been indicated in the left column; the IACV categories associated with these hypotheses are described in the right column.

TABLE 4.4. Communication Act and associated IACV categories

Hypotheses	IACV categories
1. Requests for Information	Frequencies of the acts coded as requests for information
2. Adults use more controls, tutorials	Frequencies of the various communication act categories used by the adult

4.5.5. Communication Breakdowns and Repairs
The information exchange in SLI child-caregiver conversations is often hampered by deficiencies (communication breakdowns) in the use and comprehension of pragmatic skills (Gallagher & Darnton, 1978; McTear, 1985). Compared with NLA child-caregiver conversations, caregivers spend more time in securing discourse coherence (Roth & Spekman, 1984; Blau, 1986; Prather et al., 1989).

The results of studies on communication breakdowns and repairs, relating to NLA and SLI children lead to a set of hypotheses for the present study. Compared with NLA child-caregiver interaction, SLI child-caregiver conversations are characterised by:
1. more unfinished verbal acts of the SLI child
2. a high number of parallel talk
3. more clarification requests
4. more corrections of previous acts used by the adult
5. more consecutive initiations
6. more faulty responses to preceding acts of both partners
7. more faulty initiations of both partners.

The IACV categories needed to find (counter-) evidence for these hypotheses are presented in the next table (Table 4.5.). In this table the hypotheses have been indicated in the left column; the IACV categories associated with these hypotheses are described in the right column.

TABLE 4.5. Communication Breakdown, Repairs and associated IACV categories

Hypotheses	IACV categories
1. Unfinished utterances	Total number of unfinished verbal acts of the child
2. Parallel talk	Total number of parallel talk, iniated by either the child or the adult
3. Clarification requests made by the adult	Total number of clarification requests used by the adult
4. Corrections of previous acts made by the adult	Total number of corrections of previous acts used by the adult
5. Consecutive initiations	Total number of consecutive initiations
6. Faulty partner related responses	Total number of faulty responses on previous acts of the partner for adult and child
7. Faulty initiations of adult and child	Total number of faulty initiations of adult and child

4.6. Summary

In this chapter the structural units used in IACV have been discussed. The research hypotheses, based on the literature survey (chapter 3), were formulated in the last five sections. The linguistic categories for coding the discourse (in)coherence were part of the variables used in the IACV project. The next chapter describes the 'Research Method'. The results of the study are presented in chapter 6. The hypotheses formulated in the preceding sections were:

Turn-taking
Compared with NLA children of the same age and their caregivers:
1. SLI children use more backchannelling and ellipsis
2. caregivers of the SLI children take longer turns
3. caretakers of SLI children initiate or reinitiate themes more often
4. caretakers of SLI children use more self-repetitions
5. SLI children use more imitations of preceding adult acts
6. initiative behaviour of SLI children is mostly nonverbal.

Child Adjusted Register (CAR)
Compared with NLA children of the same age, SLI children receive:
1. a language input characterised by a smaller MLU
2. a language input consisting of more (adult) self-imitations
3. a language input consisting of more (adult) requests for information.

Topic and Theme Management
Compared with NLA children of the same age, the communicative interaction between SLI children and their caregivers is characterised by:
1. more child initiations of 'introduction new themes'
2. shorter themes in general
3. shorter themes initiated by caregivers of SLI children
4. more imitations of adult acts by the child
5. more imitations of self-related acts by the adult.

Communication Functions
Compared with NLA children of the same age, SLI children:
1. use a fewer amount of requests for information
2. receive a language input with more controls and tutorials.

Communication Breakdowns and Repairs

Compared with NLA child-caregiver interaction, SLI child-caregiver conversations are characterised by:
1. more unfinished verbal acts of the SLI child
2. a high number of parallel talk
3. more clarification requests
4. more corrections of previous acts used by the adult
5. more consecutive initiations
6. more faulty responses to preceding acts of both partners
7. more faulty initiations of both partners.

CHAPTER 5

Research method

The current study tries to find differences in the use of devices for creating discourse coherence in the conversations between NLA and SLI children and their caregivers. The hypotheses for this research have been given in the preceding chapter. The present study was set up in order to find (counter) evidence for these hypotheses. As mentioned earlier this investigation is part of the IACV project (section 1.2.). Although information about the use of cohesive devices (morpho-syntactic and lexical structures) is available in several studies (Wells, 1985; McTear, 1985), this study is mainly focussing on ways in which caregivers and their NLA and SLI children create discourse coherence. The IACV project resulted in a longitudinally collected set of language data of adult and child. In this chapter an account of the research method is given.

5.1. Introduction

The IACV project* comprises a longitudinal study of the language and com-

* The IACV project has received financial support from the Dutch Praeventiefonds (projectnr. 28-1180); it started in July 1985 and was finished in May 1990.

munication development of NLA and SLI pre-school children (from 2;6 to 4;0). During the IACV project spontaneous verbal and nonverbal conversational data of adult and child were collected at two month intervals for a period of eighteen months. The ultimate result of the IACV project is a linguistic method for analysing the communicative interaction between caregivers and their SLI children. The methods for transcription and analysis in IACV are based on a standardised procedure, described in a comprehensive coding manual (Van Balkom et al., 1989; see also Appendix A). The availability of special IACV software programmes for transcription, grammatical and pragmatical analysis facilitates the work in IACV.

This chapter is not the place to discuss the working procedures of the IACV software programme. In the following sections a brief description of the research method is given.

5.2. Selection procedures

A total of eighteen dyads participated in the study; twelve dyads consisted of caregivers and their SLI children (four girls/eight boys) and six dyads consisted of caregivers and their NLA children (three girls/three boys).

In addition to the twelve SLI child-caregiver dyads another four SLI child-caregiver dyads participated as reserves. This small reserve group of children and their caregivers is needed in order to deal with unexpected drop out of participants (cancelling participation, illness, divorce which leads to intermediate change of the caregiver; moving house). All children were selected according to the following criteria (see also section 1.4.):
- chronological age
- family background
- language comprehension and production
- hearing abilities
- psycho-social functioning
- medical history.

All SLI children were referred to the regional Speech and Language Training Centre in Hoensbroek. The NLA children came from local play and pre-schools. The SLI children were selected with the formal intervention and cooperation of the Speech and Language Training Centre. The selection of NLA children was organised by directly contacting the parents of the NLA children in local play- and pre-schools.

5.2.1. Chronological age and family background
As mentioned earlier, the IACV project aimed at studying the language and communication development of SLI and NLA children during their most sensitive period for acquiring and experiencing communicative abilities (at the age of 2;6 to 4;0). The socio-educational level (based on the educational level and the actual profession or employment of the parents) was recorded, but did not count as a selection criterion.

In addition to the selection criteria for SLI children, their parents had to meet the following requirements:
- speaking Dutch as native language or
- speaking Dutch as the official language and having the regional dialect as the native language
- no history of hearing impairments, language and learning disorders
- complete family constitution.

All parents of SLI and NLA children involved in the IACV project were acquired the regional dialect as their native language and used Dutch as the official language.

The strict selection criteria for SLI caused problems in trying to rapidly complete a group of SLI children. Most children initially diagnosed as SLI appeared to have additional problems (e.g. frequent hearing loss due to 'otitis media', epilepsy) or one of the parents had or has had language problems (such as stuttering, reading and writing difficulties). For that reason the age criterion was modified. All children with language development disorders were preselected for IACV if their age fell within the range from 2;6 to 3;0. The SLI children were preselected in cooperation with the Speech and Language Training Centre for all children in the agegroup of 2;6-3;0. The speech therapist of the IACV research team was involved in contacting the parents of NLA children. The children, registered at the Speech and Language Training Centre as developmentally language impaired, were screened on the criteria for selection stated above.

Before starting the testing and selection procedure, the Speech and Language Training Centre informed the parents of the preselected language impaired children about the project and asked their approval for transferring the necessary information of their child to the IACV research team. The speech therapist of the IACV research team used a similar approach when contacting the parents of NLA children.

After the parents formally accepted their participation in the IACV research project a home-visit was arranged. This home-visit intended to inform the parents about the research project and their involvement in the longitudinal language study. In addition to that a personal interview was carried out with one parent of each child prior to the actual participation in the study. In this interview the speech therapist obtained information regarding the child's birth, medical history and the child's early social and communicative skills. During that first home-visit the child was also tested for language comprehension. The interview was based on an existing questionnaire and checklist at the Speech and Hearing Centre. The questionnaire and checklist concerning the social and communicative skills of the children from birth to five years old was based on a Dutch version of the 'Early Social Communicative Scales (ESCS)' from Seibert and Hogan (1982, 1984). The ESCS interviews also provided information needed for the selection of play materials for the child at the first free play session (see Appendix G).

5.2.2. Subjects

All SLI children were screened according to the criteria described above. The NLA children were tested on language comprehension and language production only at the start of the study. The SLI children were mostly screened by one or more medical specialists (a paediatrician, an Ear, Nose and Throat (E.N.T.) specialist or a neurologist). All children (in both the SLI and NLA group) had uneventful case histories with respect to peri- and postnatal, neurological and socio-emotional problems. As can be seen from the Tables 5.1. and 5.2., the SLI children ranged in age from 2;7 to 3;2 with a mean age of 2;11 and the NLA children ranged in age from 2;6 to 3;1 with a mean age of 2;10. The delay in language comprehension of the SLI children ranged from 0 to 9 months, as indicated on the Revised Edition of the Reynell Developmental Language Scales (RDLS) and the Peabody Picture Vocabulary Test (PPVT) (section 5.2.3.). Most of the SLI children were referred to the Speech and Hearing Centre because they were judged as having significant expressive language problems. The mean percentage of grammatically not analysable utterances (in GRAMAT) of the SLI children is 57% (Range 40%-87%, SD 14%) and of the NLA children 34% (Range 24%-40%, SD 8%) at session 1. The remaining grammatically analysable utterances (in GRAMAT) of the SLI children were at Clause level 1 and 2 and for the NLA children at Clause level 3 and 4. The delay in language production of the SLI children (based on MLU* measures at beginning of the study, session 1) ranged from 10 to 20 months (with a mean delay of 14 months). The SLI children were all enrolled in a language remediation programme at the start of the study at the Speech and Hearing Centre. These programmes primarily intended to stimulate the child's development of expressive skills and did not involve parental instruction or a guidance plan, although the parents were encouraged to stay and observe their children and discuss progress with the speech therapist.

* The language production age was primarily based on MLU-language associations, published in Wells (1985). There are no comparable data for the Dutch language.

The average MLU for the SLI children was 1.62 (range 1.02-2.57, SD 0.47) and for the NLA children was 3.62 (range 2.89-4.66, SD 0.72). The average MLU-equivalent language production age for the SLI children is about 1;6 and for the NLA children about 3;3. All NLA children demonstrated age appropriate or superior language performance. The tables 5.1 and 5.2 present the data for the SLI and NLA children at session 1. The tables 5.3. and 5.4. present the data for the SLI and NLA children at session 9 (final session). The data of session 9, presented in the tables 5.3. and 5.4. indicate the way in which the SLI and NLA children developed in their language acquisition during the research period. The SLI children on average still show more grammatically unanalysable utterances (45%) and a smaller average MLU (3.26) than the NLA children (34% and 4.62 respectively).

At session 9 most of the grammatically analysable utterances of the SLI children were at Clause level 1, 2 and 3 and for the NLA children at Clause level 3, 4 and 5. Although language comprehension seems to be at age appropriate levels for the SLI children (as indicated by RLDS and PPVT scores), they still seem to have difficulties in achieving an age appropriate language production performance (as indicated by GRAMAT).

TABLE 5.1 Subject Information SLI Children (N=12), at session 1

(1)	(2)	(3)	(4)	(5)	(6)	(7)
1	Linda	f	3;1	-6	2.25	43%
2	Saskia	f	2;10	0	1.35	68%
3	Rianne	f	2;9	-4	1.56	41%
4	Davy	m	3;0	-3	1.56	57%
5	Lisette	f	3;1	0	1.97	67%
6	Geronimo	m	2;7	0	1.02	87%
7	Marcel	m	3;1	0	2.30	40%
8	Bjorn	m	3;2	-1	1.79	47%
9	Sebastian	m	2;10	-9	1.13	72%
10	Johny	m	3;2	-8	2.57	65%
11	Christiaan	m	3;2	?	1.61	47%
12	Remco	m	3;0	0	1.59	52%
Mean			2;11	-3	1.62	57%

TABLE 5.2. Subject Information NLA Children, at session 1

(1)	(2)	(3)	(4)	(5)	(6)	(7)
13	Rachel	f	2;6	+14	2.89	37%
14	Bram	m	3;0	+3	4.04	32%
15	Tim	m	2;8	+36	4.02	40%
16	Michel	m	3;0	+18	3.04	37%
17	Rosanna	f	3;0	+26	3.07	33%
18	Chantalle	f	3;1	+13	4.66	24%
Mean			2;10	+18	3.62	34%

(1) Dyad Number (DN)
(2) Name of Child
(3) Gender
(4) Chronological Age (years;months)
(5) Language Reception in months (RDLS)
(6) MLU (measured in syllables)
(7) Grammatically not analysable (% of verbal acts child)

? Not testable
0 No Delay*
- Delay*
+ Advance*
m masculine
f female

* Compared to (4) Chronological age

TABLE 5.3. Subject Information SLI Children (N=12), at session 9

(1)	(2)	(3)	(4)	(5)	(6)	(7)
1	Linda	f	4;9	+9	2.27	51%
2	Saskia	f	4;6	+14	2.90	38%
3	Rianne	f	4;4	+9	1.65	31%
4	Davy	m	4;8	+16	3.83	39%
5	Lisette	f	4;9	+3	3.70	40%
6	Geronimo	m	4;2	+8	2.24	47%
7	Marcel	m	4;9	+7	4.80	47%
8	Bjorn	m	4;10	+2	3.18	42%
9	Sebastian	m	4;5	+6	2.32	50%
10	Johny	m	4;9	+2	3.21	53%
11	Christiaan	m	4;8	+2	5.07	54%
12	Remco	m	4;6	+4	3.95	44%
Mean			4;7	+7	3.26	45%

TABLE 5.4. Subject Information NLA Children, at session 9

(1)	(2)	(3)	(4)	(5)	(6)	(7)
13	Rachel	f	4;1	—*	4.19	34%
14	Bram	m	4;7	—	5.36	32%
15	Tim	m	4;5	—	4.91	35%
16	Michel	m	4;8	—	4.10	40%
17	Rosanna	f	4;7	—	3.30	42%
18	Chantalle	f	4;8	—	4.36	23%
Mean			4;6		4.62	34%

(1) Dyad Number (DN) ? *Not testable*
(2) Name of Child 0 *No Delay***
(3) Gender - *Delay***
(4) Chronological Age (years;months) + *Advance***
(5) Language Reception in months (RDLS) m *masculine*
(6) MLU (measured in syllables) f *female*
(7) Grammatically not analysable (% of verbal acts child)

** Compared to (4) Chronological age

* All NLA children scored far above their chronological age on the RLDS and PPVT

5.2.3. Language production and language comprehension

All children (NLA and SLI) were tested on language comprehension and language production at home during the first home-visit (in the case of NLA children) or at the Speech and Language Training Centre (for the children with suspected SLI). On the basis of the results of these tests, the available medical data and the parent-questionnaire (interview), the speech therapist classified the child as specific language impaired or acquiring language normally.

The language tests used were:
- RDLS Revised Edition/ Expressive and Verbal Comprehension Scales (Verbal Comprehension Scale A) (Reynell, 1974)
- The experimental Dutch version of the Peabody Picture Vocabulary Test (PPVT) (Dunn, 1976)
- Early Social Communicative Scales (ESCS) (Seibert & Hogan, 1982).

The raw test scores of PPVT and RLDS were given as equivalent age scores, indicating the delay or advance in language comprehension age (see the Tables 5.1., 5.2. and 5.3.). For all NLA and SLI children ultimately selected these tests were used at the first home-visit, the second home-visit (after one year of participation in the IACV project) and the third home-visit at the end of the research period (after eighteen months).

The SLI children were involved in a further assessment protocol in order to screen additional or accompanying disorders. The children whose language impairment was primarily characterised as articulation deficiency or whose language impairment was connected with other disorders (e.g. chronic middle-ear problems ('otitis media'), severe hearing impairment, brain lesion, mental retardation) should not be classified as SLI and were not selected for participation in IACV. The criteria for selecting SLI children and their caregivers are discussed in the following sections.

5.2.4. Hearing abilities

The children with suspected SLI were tested for possible hearing problems by the audiologist and speech therapist of the Speech and Language Training Centre (in Hoensbroek). All children had to go through audiometric screening procedures for frequencies from 500 to 2000 Hz at 25 db. The definition and selection of SLI was not influenced by suffering from temporary loss of hearing caused by incidental otitis media or a cold; these SLI children were included as potential subjects for the IACV project.

5.2.5. Psycho-social functioning

If there was evidence of additional cognitive and intelligence deficiencies (based on psychological testing), the language impaired child was excluded from IACV participation. These children were diagnosed as autistic, socially deprived, dysphatic (e.g. brain lesions caused by traffic accidents), mentally retarded or Minimally Brain Damaged (e.g. caused by anoxia in the perinatal stage of birth). In most cases the psychologist of the Speech and Language Training Centre conducted the psycho-social testing.

5.2.6. Medical history

All information concerning specific medical etiologies (e.g. E(ar) N(ose) T(hroat) diseases) also led to the exclusion of participation in the IACV project. As mentioned at the beginning of this section the strict criteria for selecting SLI children caused problems in soon completing the research group. Of the total number of children referred with presupposed SLI to the centre in Hoensbroek (53), only 10 could actually be classified as SLI in 1985. The additional SLI children who completed the IACV research (and reserve) group came from other regional hospitals with a speech and hearing department.

5.3. Situation and Equipment

Each adult-child pair was invited to visit the Speech and Language Training Centre in Hoensbroek for a free-play session at regular intervals of two months during the total period of eighteen months. Each session was scheduled for thirty minutes. The adult (mostly the mother and in three cases the father) was asked to play with his or her child in exactly the same way as at home.

Caregiver and child played in a room with special child friendly furniture and play materials. The amount and kind of play materials varied at each session and consisted of three different objects. The toys were appropriate for children at age two to five years and were selected beforehand. They consisted of objects stimulating mutual conversation and activities (e.g. animalfarm, Fisher-Price Zoo and garage; mechanical toys; game of skittles; a medical kit; a telephone). The selection of play materials commonly used in eliciting social-communicative behaviour in toddlers is important. Books, wall posters and pictures were not used, because they elicit stereotype and non-spontaneous communication. Substitution of objects for the subsequent play sessions is needed, because the child's interest changes over a period of eighteen months. In all these cases substitution was made according to the child's interest and comments of the parent. Parent and child played in a room without the presence of the observers. The room was fitted with a one-way mirror screen and specific facilities for connecting microphones with audio- and video-equipment in the room behind the one-way mirror screen. The playroom was a quiet room, likely to be free from outside distractions or interruptions. The room was furnished in a simple way with a child-size wooden table and chairs. Adults and children usually preferred to sit or kneel on the floor. The walls were decorated with Sesame Street posters and child drawings. Three different kinds of toys were grouped on the floor in the middle of the room. A normal size table was placed in front of the one-way screen, preventing the child from playing in an area out of reach of the video-camera in the other room.

The room adjacent to the playroom was used for observation, a video-camera (type colour Philips VK 4002) was installed behind a one-way mirror screen. In the playroom a microphone (type Sennheiser ME 20) was used and an additional audiotape-recorder (type Philips D 9610), connected with the same type of microphone, was placed out of sight. The video-camera was connected with a UMATIC colour-recorder and player (type JVC CR 6650E) and a colour monitor (type BARCO CR 2032).

At each session two people were present, one to operate the video equipment and one to comment on the activities. After each session child and parent were asked to review the playsession or parts of it; their comments were written down in the session report.

5.3.1. *Transcription*

The observers took a random selection of an uninterrupted five-minute sample from each adult-child recording. The recordings were coded in the video-studio in order to indicate the session number and time (in seconds). The five-minute sessions were transcribed completely with equal emphasis on verbal and nonverbal acts of both partners. Paraphrases, interpretations, situational and context information were given in addition to the described language behaviour.

The UMATIC video-recorder, video-player and monitor were used for transcription. The audio-recordings served only as an extra check for speech which was difficult to understand. For purposes of efficient and accurate trancription, a (self-made) foot-switch for operating the video-player, an amplifier and earphones were used. In order to have a reliable and standardised transcription, a comprehensive manual for IACV transcription was devised and evaluated through several pilot studies (see also Ochs, 1979). The transcription manual is part of the IACV coding manual for transcription and analysis (Van Balkom et al, 1989; see also Appendices A and B).

One minute segments of five randomly chosen transcripts of different children and sessions were transcribed independently by two other transcribers. The transcript segments were checked by comparing them afterwards with the original transcripts. We did not statistically test the reliability of the transcript segments. A close and careful examination of the segments revealed no differences in the segmentation of verbal and nonverbal acts. There were some problems in indicating simultaneously occurring acts of both partners, in specifying situational and contextual data in the transcripts and in the correct transcription of verbal acts spoken in the regional dialect. These problems and disagreements were resolved by making final transcripts in conference based on the videotaped material.

5.3.2. IACV Analysis Procedure and Coding Reliability

The transcriptions were analysed using the specifically designed IACV programmes for morpho-syntax and pragmatics. The morpho-syntax analysis procedure is divided into two separate parts. The child's verbal acts are analysed according to GRAMAT (Bol & Kuiken, 1989). Adult verbal acts are described on a set of descriptive syntactic categories, e.g. MLU, sentence complexity, the main verb used, grammatical correctness and Child Adjusted Register (CAR). The pragmatic analysis component is specifically designed for characterising verbal and nonverbal acts of both partners on the basis of a set of categories such as: initiations, responses, comments, overlap relations (parallel talk, nonverbal-verbal overlaps), the kind of relation with the preceding verbal or nonverbal act, the intended direction (self, partner or other), coherence relations, imitations and topic structure. Within the pragmatic analysis, verbal and nonverbal acts of both partners are grouped together in subsequences, sharing the same thematic content (a theme).

A 'theme' is defined as a group of verbal and nonverbal acts between two initiations. Each subsequence is characterised in terms of its main communicative function. The acts in a subsequence are classified in terms of communication functions belonging to the assigned main communicative function of the subsequence (see example 8, section 4.2.6.). The IACV grammatical analysis starts with the first verbal act in the transcript. The IACV pragmatic analysis starts with the first verbal act or nonverbal act in the transcript.

The sampling procedure (5-minutes sample transcripts) causes a variation in the size of each transcript (number of verbal acts and nonverbal acts of caregiver and child). Each transcript is analysed completely.

The reliability of the IACV pragmatic coding is established by having two persons* independently review 12% (at least 10% is required according to Cohen, 1960) of all transcripts and independently classify the pragmatic categories for adult and child acts. The percentage of agreement between the two coders is calculated using Cohen's 'Kappa' (1960) based on 12% of the transcripts (eighteen transcripts of which twelve came from SLI dyads and six from NLA dyads). This gives the following overall result for all IACV Prag-

---

* The two persons involved in the reliability study were the original coders.

matic Categories: 0.78 (95% confidence interval 0.74, 0.83). Cohen's 'Kappa' is calculated for all the IACV Pragmatic categories separately. The results of these calculations are given in Appendix D. Because Cohen's Kappa is a very conservative measure, the obtained findings for the various IACV Pragmatic Categories suggest acceptable interjudge reliability for the analyses. Common errors of reliability in the study conducted were interchanges between for example:
- 'initiations, elaborations of theme' (defined in section 4.3.1) and 'comments' (defined in section 4.3.2.)
- 'partner-related acts' and 'self-related acts' (defined in section 4.3.4.)
- 'control function: requests' and 'expressive function: query state or attitude' (defined in Appendix C, section 3.1.-3.3. and section 4.3.)

5.3.3. Compiling the data

After completing all IACV analyses for the SLI and NLA child-caregiver dyads, the data were compiled for statistical processing (SPSS, Statistical Package for the Social Sciences) in order to fulfil the objective of our study and to gather (counter-) evidence for the hypotheses stated. The IACV categories used to collect the language data for discussing the hypotheses have been described in the final sections of chapter 4. The results of the study conducted are presented in the next chapter.

In order to present the results efficiently, the data of the SLI- and NLA child-caregiver dyads are presented as two separate groups: the group-data of the SLI child-caregiver dyads and the group-data of the NLA child-caregiver dyads. We are mainly interested in group differences between the SLI and NLA dyads. All IACV pragmatic variables and most IACV morpho-syntax variables are nominal. The statistical analysis in the study at issue deals primarily with frequency counts of the IACV pragmatic variables. A more detailed description and discussion of the individual characteristics and intra- and interindividual differences of the children studied will be given in a separate publication. Because of the nature of this study we will not discuss morpho-syntax and its relation to pragmatics deeply, although the data are available in IACV. In the next chapter frequency distributions, percentages and analysis of variance are used to evaluate the earlier hypothesised differences within and between the means of frequencies of both groups.

5.4. Summary

This chapter has given an account of the IACV research design from which the data for this discourse coherence study are derived. There are some specific problems with longitudinal language studies, which are not discussed in this chapter, such as external influences which are difficult to handle and can threaten the validity of the study (e.g. biased selection of subjects, maturation in language development, changes in family constitution; see Cook & Campbell, 1979; Van Balkom & Heim, 1990). We have tried to tackle these threats by carefully selecting the subjects and gathering as many data as possible of the family background during three home-visits, by achieving as much consistency as possible in our organisation of free play sessions, transcription and analyses procedures and by using a small reserve group of SLI children and their caregivers. The reliability of the IACV pragmatic coding was established by reviewing 12% of all transcripts and using Cohen's 'Kappa' (1960).

After a formal description of the research design in this chapter, the next chapter presents the data gathered from the IACV project in order to find (counter) evidence for the hypotheses stated in chapter 4.

CHAPTER 6

Results

This chapter presents the results of the discourse coherence study at issue. The acceptance or rejection of the hypotheses formulated in chapter 4 is discussed in the light of the data gathered from this study. The data are presented in graphs and tables. The data displayed in the graphs are given in percentages of all acts (per session), except for the Mean Length of Utterance (MLU), the Mean Length of Turn (MLT) and the Mean Length of Subsequence (MLS), which are given as numerical scores. Two kinds of tables with statistics are used in this chapter:
- frequency-tables presenting the group-means and standard-deviation of the scores of all caregivers and/or children per session, also indicating the overall means and standard deviation (all sessions and all caregivers/ children). These tables are sequentially numbered ('T+number') and presented in Appendix F
- tables presenting the results of the 'F-Statistics' (MANOVA's: Analysis of Variances for group and session means). These tables are presented at the start of each main section in this chapter.

Two kinds of MANOVA's were used:
- MANOVA's for so-called 'Session-Effect' (indicating the significance of differences between the scores of caregivers or children per session)
- MANOVA's for so-called 'Between Groups Effect' (indicating the significance of differences between the caregivers or the children from the SLI Dyads and the NLA Dyads).

In chapter 4 five groups of hypotheses were discussed. The results of the two MANOVA's for each group of hypotheses are summarised in one table. In the MANOVA Tables the results of the tests of significance for the 'Session Effect' and the 'Between Groups Effect' are indicated for three probability (p) levels: $p \leq 0.001$ (indicated with ***), $p \leq 0.01$ (indicated with **) and $p \leq 0.05$ (indicated with *).

Chapters 6 and 7 will not extensively address the issue of individual differences in subject scores. Although these individual differences offer very important and interesting information, the context of this study does not permit in-depth discussions about individual and intra-subject differences. This issue will be addressed in a forthcoming report of the overall results from the IACV project.

The actual data about the total number of acts used by the children and their caregivers are given in the next section. The percentage-scores in the graphs are based on these totals.

6.1. Total Number of Acts used by Adult and Child

The percentage-scores of the categories studied are based on the total number of verbal and nonverbal acts used by adult and child for each session. This was done mainly because we are interested in the differences in proportions of communicative behaviours between the conversational partners.

The Tables F1a,b (Appendix F) present the nine session totals, means and standard deviations of the total number of acts, the total number of verbal acts and the total number of nonverbal acts used in each session for the NLA Dyads and the SLI dyads (see Figure 6.1a. "Total Number of Acts used in the NLA Dyads" and Figure 6.1b. "Total Number of Acts used in the SLI Dyads").

Table 6.1. presents the results of the F-statistics for 'Session Effect' and 'Between Groups Effect'.

TABLE 6.1. Results of MANOVA's for Session and Between Groups Effect: Total Acts, Verbal and Nonverbal Acts

	SESSION EFFECT		GROUPS EFFECT
Total Number of Acts used	SLI Sign.of F (df 88/df 8)	NLA Sign. of F (df 40/df 8)	Sign. of F (df 16/df 1)
Overall-total of Acts	0.427	0.747	0.096
Total Acts Adult	—	—	0.300
Verbal Acts Adult	0.547	0.432	0.978
Nonverbal Acts Adult	0.320	0.945	0.110
Total Acts Child	—	—	0.036*
Verbal Acts Child	0.346	0.557	0.111
Nonverbal Acts Child	0.049*	0.890	0.036*

* $p \leq 0.05$
** $p \leq 0.01$
*** $p \leq 0.001$

Figure 6.1a. Verbal and nonverbal acts NLA dyads

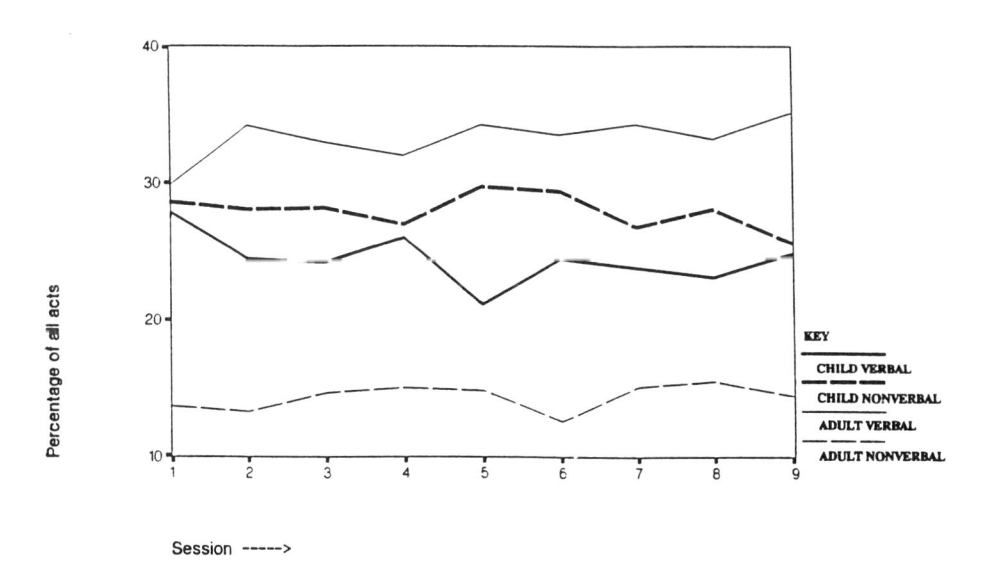

Figure 6.1b. Verbal and nonverbal acts SLI dyads

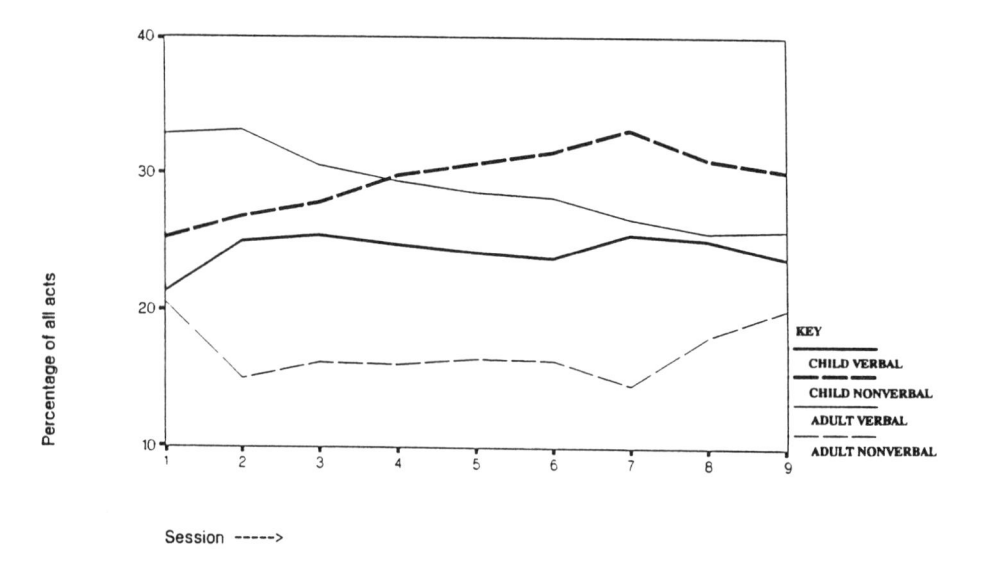

The NLA as well as the SLI dyads show no significant differences in the session means. There is a significant 'Between Groups Effect' for the total number of acts used in the NLA and the SLI dyads and a significant 'Between Groups Effect' for the total number of acts used by the NLA and the SLI children. There is also a (nearly) significant 'Between Groups Effect' for the nonverbal acts of the NLA and the SLI children, indicating that the SLI children use significantly more nonverbal acts than the NLA children. The 'Between Groups Effect' for the children's acts seems to be the result of the increasing number of nonverbal acts of the SLI children. The number of nonverbal acts increases from session one to nine for the SLI children (nearly significant) and remains stable for the NLA children.

In the NLA dyads more than 50% of the average number of acts remains verbal (Figure 6.1a.) In the SLI dyads the average number of verbal acts decreases from about 60% in the first three session to 40% from session 5 to session 9, mainly caused by the decrease in the average number of verbal acts of the caregivers (Figure 6.1b.).

6.2. Turn-taking

TABLE 6.2. Results of MANOVA's for Session and Between Groups Effect: Turn-Taking

	SESSION EFFECT		GROUPS EFFECT
Hypotheses Turn-taking	SLI Sign.of F (df 88/df 8)	NLA Sign. of F (df 40/df 8)	Sign. of F (df 16/df 1)
Backchannels Child	0.691	0.306	**0.019***
Adult	**0.000***	0.641	0.462
Ellipsis Child	**0.004****	0.958	0.112
Mean Turn Length Adult	0.353	0.492	0.182
Initiations Adult	**0.000***	0.949	0.050
Reinitations Adult	**0.000***	0.160	**0.038***
Elaborations Adult	**0.000***	0.956	0.090
Imitations, self-related, Adult	**0.000***	0.612	**0.002****
Imitations, partner-related, Child	0.206	0.720	0.720
Nonverbal Initiations Child	**0.041***	0.997	**0.021***

* $p \leq 0.05$
** $p \leq 0.01$
*** $p \leq 0.001$

6.2.1. Backchannelling
Between Groups Effect

Table F2 (Appendix F) shows that backchannels are used more often by the caregivers of both groups than by their children. Table 6.2. indicates that there are no significant differences in the use of the backchannels between the two groups of caregivers. The SLI children use significantly more backchannels than the NLA children.

Session Effect

The SLI dyads show a significant 'Session Effect' of caregivers' backchannels, primarily as the result of an increase in backchannels from session 1 to 5 and from 6 to 9, with a dip at session 6 (mainly due to a decrease in frequency of backchannels used by the caregivers in DN 2, DN 3 and DN 4). The pattern of backchannel frequencies of the adults in both groups is irregular with several dips and peaks, due to varying individual frequencies of backchannels. A similar (irregular) pattern is seen for the frequencies of backchannels used by the NLA children (see Figures 6.2a and 6.2b). The frequencies of the children's backchannels in the NLA and in the SLI dyads do not lead to significant 'Session Effect' (Table 6.2.). The SLI children show an increase in backchannels from session 1 to 9, the NLA children do not show this pattern.

Figure 6.2a. Backchannels in NLA dyads

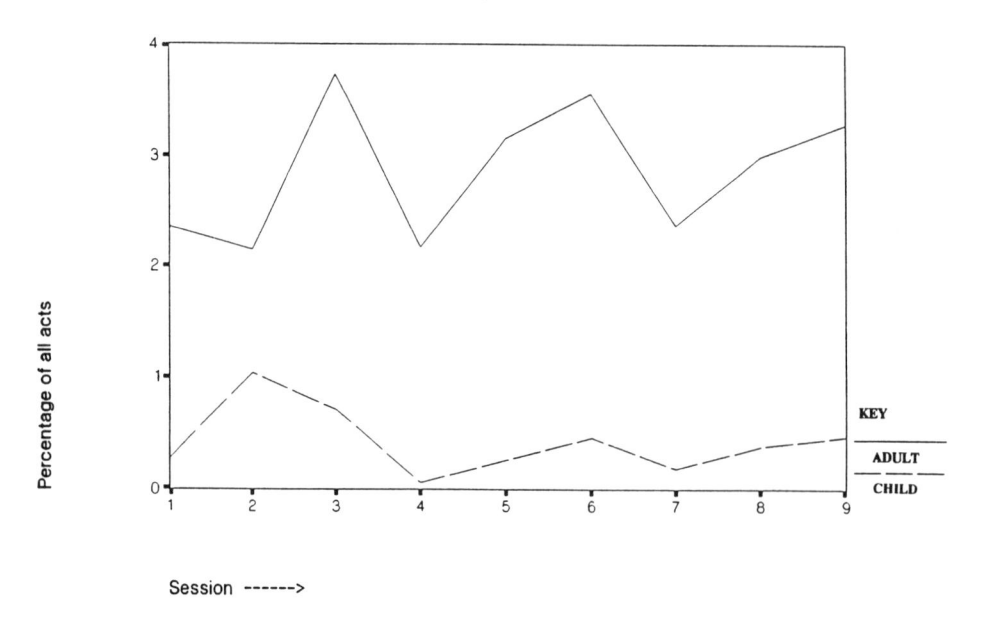

Figure 6.2b. Backchannels in SLI dyads

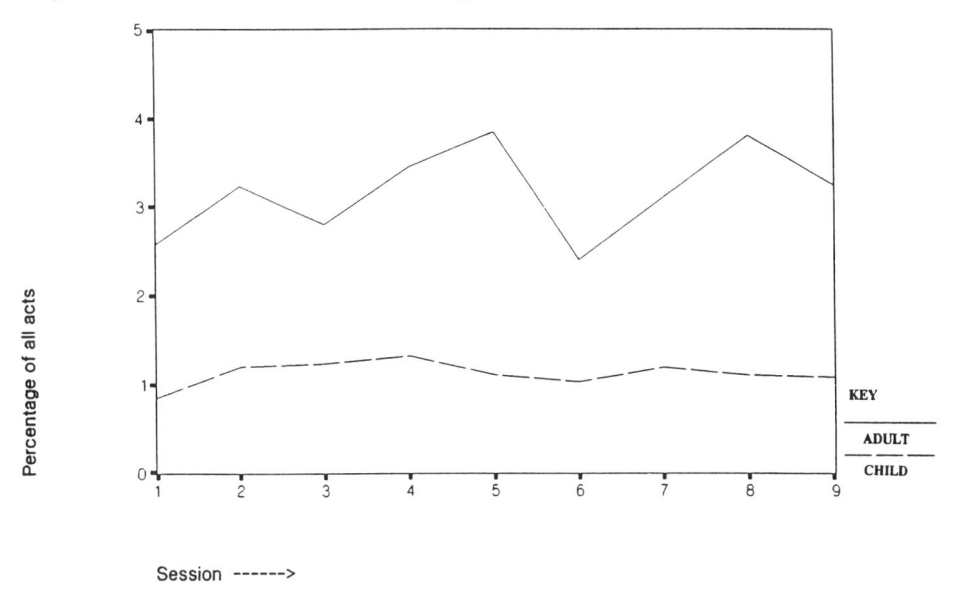

6.2.2. Ellipsis
Between Groups Effect
The Tables F2 (Appendix F) and 6.2. indicate that the SLI and the NLA children in this study do not differ significantly in their use of ellipsis.

Session Effect
The SLI and the NLA children only differ in 'Session Effect'. The SLI children show a significant 'Session Effect', the NLA children do not. The SLI children's frequency pattern of ellipsis is irregular compared with the frequency pattern of ellipsis of the NLA children.

6.2.3. Mean Length of Turn (MLT)
Between Groups Effect
The Tables F2 (Appendix F) and 6.2. indicate that the NLA and the SLI children and their caregivers do not differ significantly in MLT.

Session Effect
The MLT remains small and constant during all sessions (about 2.5 acts per turn on average). The Figures 6.4a,b,c,d illustrate the similarities in MLT.

Figure 6.3a. Ellipsis in NLA dyads

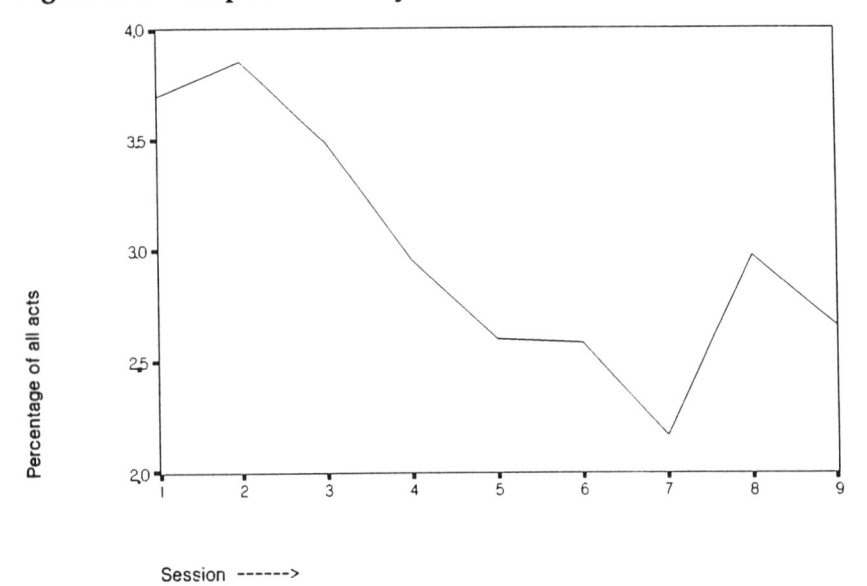

Figure 6.3b. Ellipsis in SLI dyads

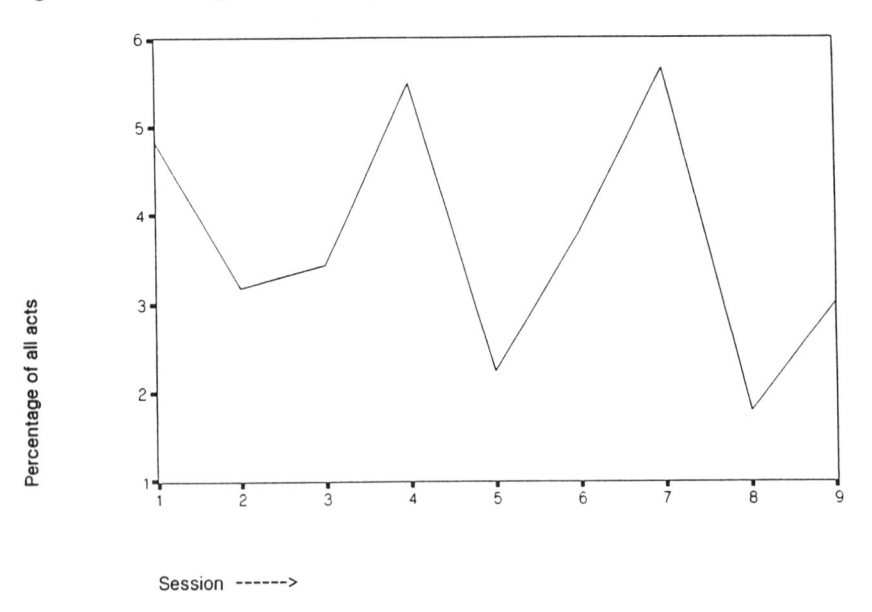

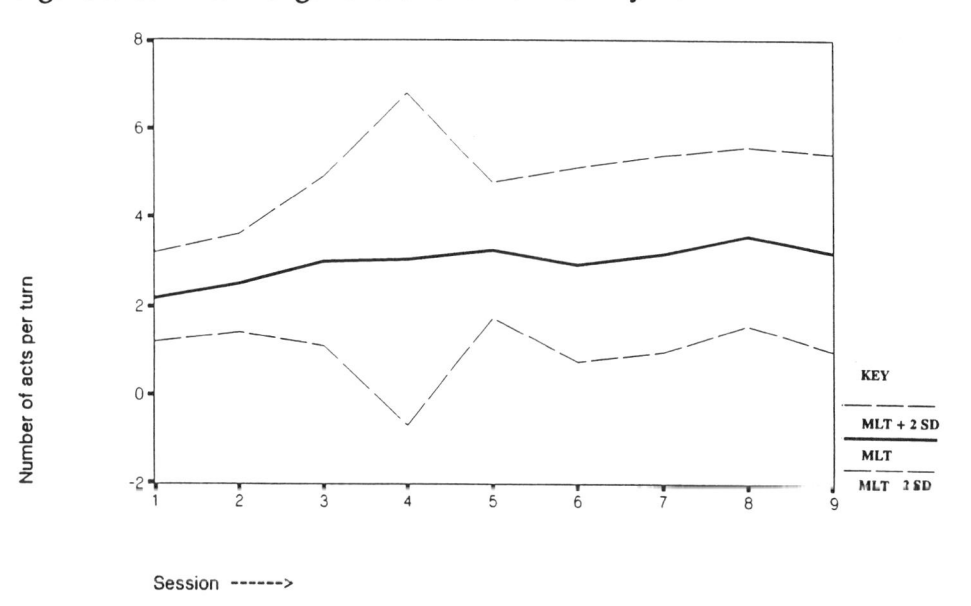

Figure 6.4a. Mean length of turn of adult in SLI dyads

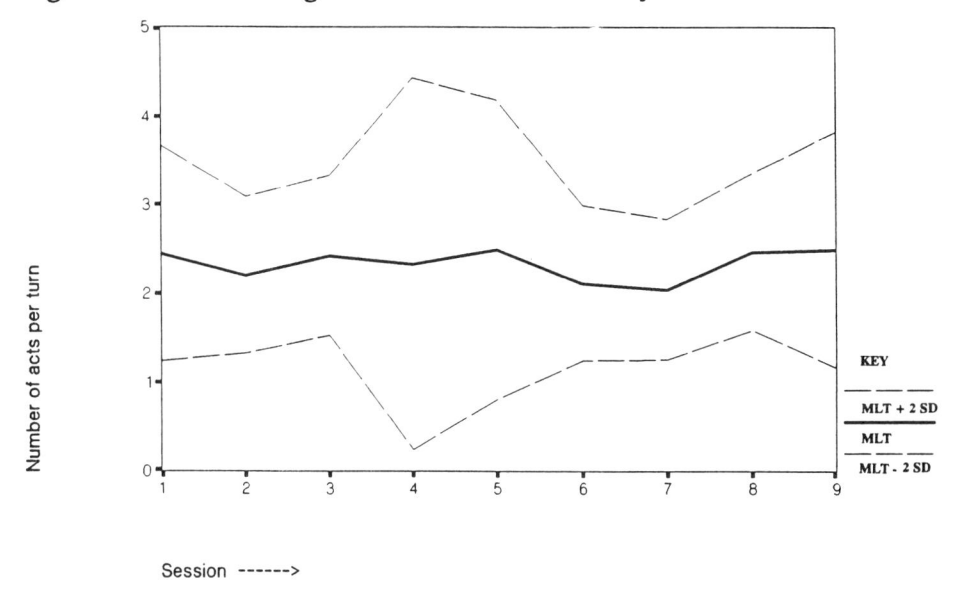

Figure 6.4b. Mean length of turn of child in SLI dyads

Figure 6.4c. Mean length of turn of adult in NLA dyads

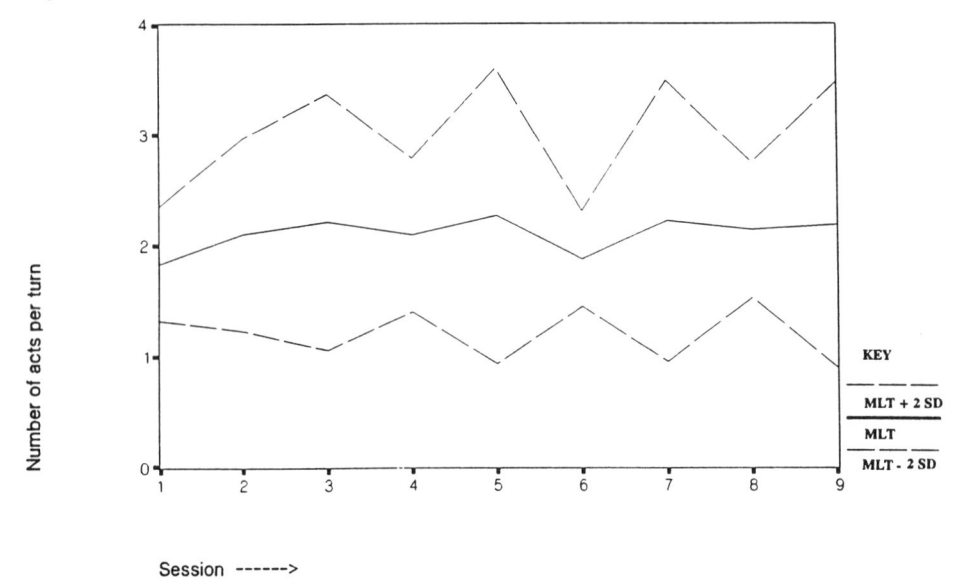

Figure 6.4d. Mean length of turn of child in NLA dyads

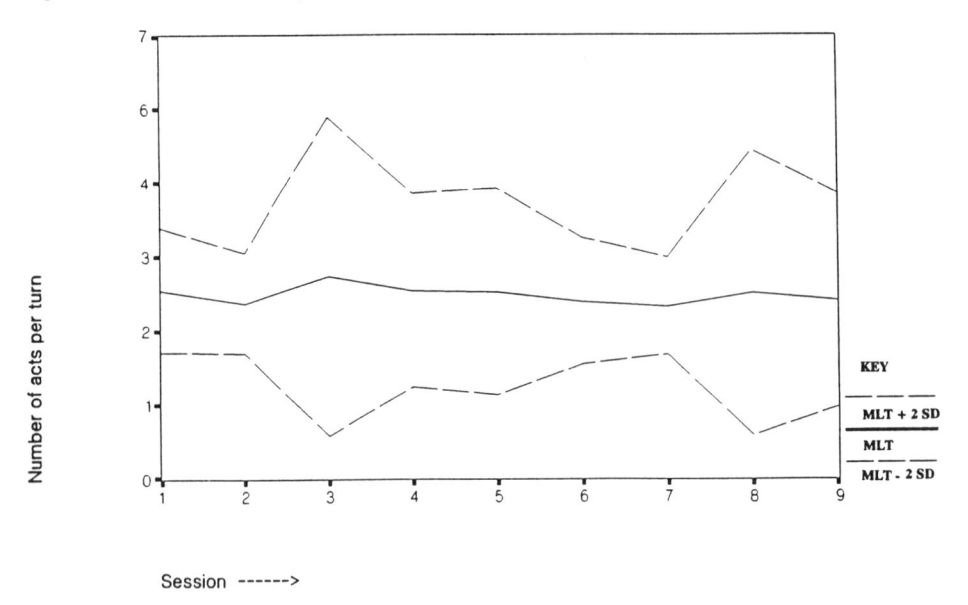

Generally the average of the MLT for the caregivers in the NLA dyads increases from session 1 to 9. The average of the MLT for the caregivers of the SLI dyads remains the same. The average of the MLT for the SLI children tends to increase from session 1 to 9; the average of the MLT for the NLA children remains the same.

6.2.4. Initiations Adult
Between Groups Effect
The Tables F2 (Appendix F) and 6.2. indicate that the caregivers of the SLI and of the NLA children tend to differ in the number of initiations used. The caregivers of the SLI children tend to use more initiations than the caregivers of the NLA children.

Session Effect
The number of initiations used by the caregivers of the SLI children decreases significantly from session 1 to 9 (Figure 6.5b.). The number of initiations used by the caregivers of the NLA children tends to increase (not significantly) from session 6 to 9 (Figure 6.5a).

6.2.5. Reinitiations Adult
Between Groups Effect
The Tables F2 (Appendix F) and 6.2. indicate a significant 'Between Groups Effect' for the caregivers' reinitiations in the NLA and the SLI dyads. The majority of all initiations used by the caregivers of the SLI and the NLA children are elaborations of themes introduced earlier. A rather small part of the initiations used are reinitiations or reintroductions of themes.

Session Effect
The Tables F2 (Appendix F) and 6.2. indicate a significant 'Session Effect' for caregivers' reinitiations in the SLI dyads. The number of caregivers' reinitiations in the NLA dyads tends to increase (not significantly) from session 1 to 9 and decreases significantly in the SLI dyads from session 1 to 9 (see also Figures 6.5a,b).

Figure 6.5a. Initiations in NLA dyads

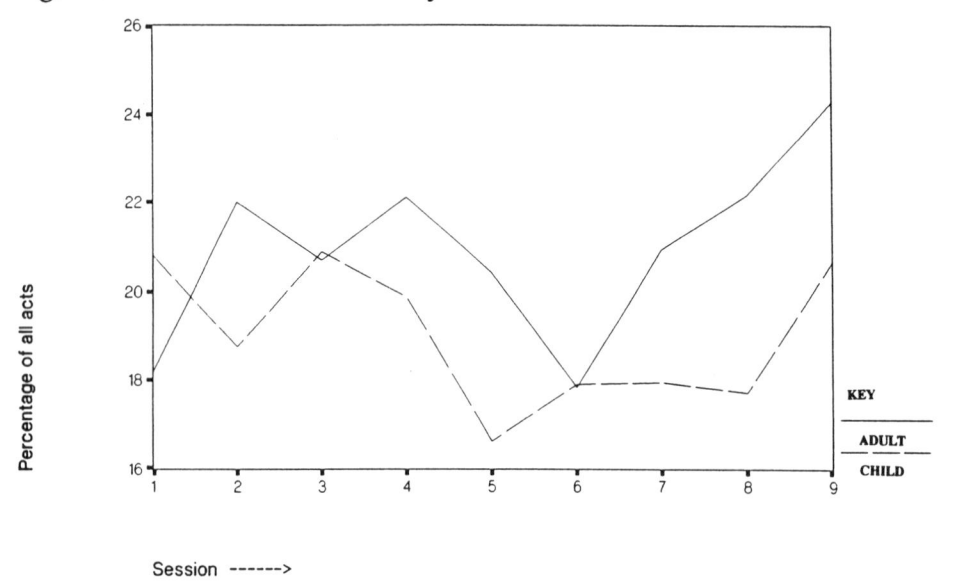

Figure 6.5b. Initiations in SLI dyads

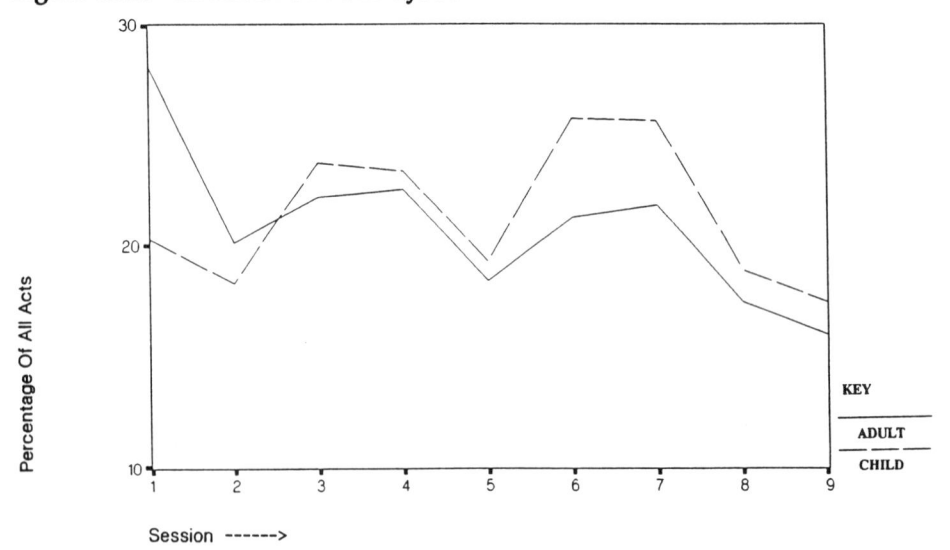

6.2.6. Self-related Imitations Adult
Between Groups Effect

The Tables F2 (Appendix F) and 6.2. indicate a significant 'Between Groups Effect' for caregivers' self-related imitations in the NLA and in the SLI dyads. Generally the caretakers in the SLI dyads use a higher number of self-related imitations than the caretakers in the NLA dyads.

Figure 6.5c. Initiations, reinitiations and elaborations adult in NLA dyads

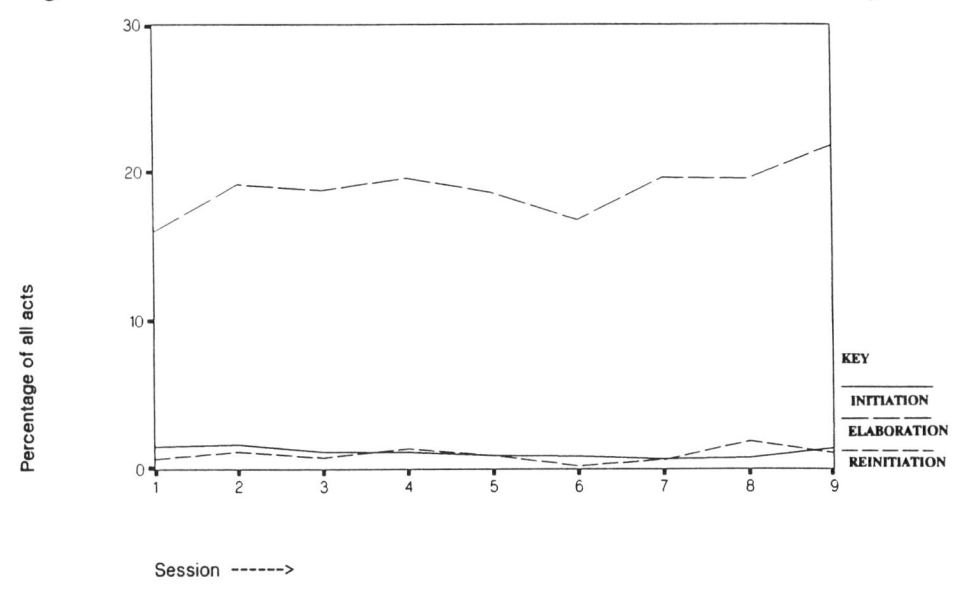

Session Effect
The Tables F2 (Appendix F) and 6.2. indicate a significant 'Session Effect' for caregivers' self-related imitations in the SLI dyads. That 'Session Effect' is mainly accomplished by a significant decrease in the caregivers' self-related imitations in the SLI dyads from session 1 to 9.

6.2.7. Partner-related Imitations Child
Between Groups Effect
The Tables F2 (Appendix F) and 6.2. indicate that there are no significant differences in frequencies of partner-related imitations used by the NLA and by the SLI children.

Figure 6.6.a. Self-related imitations of adult in NLA dyads

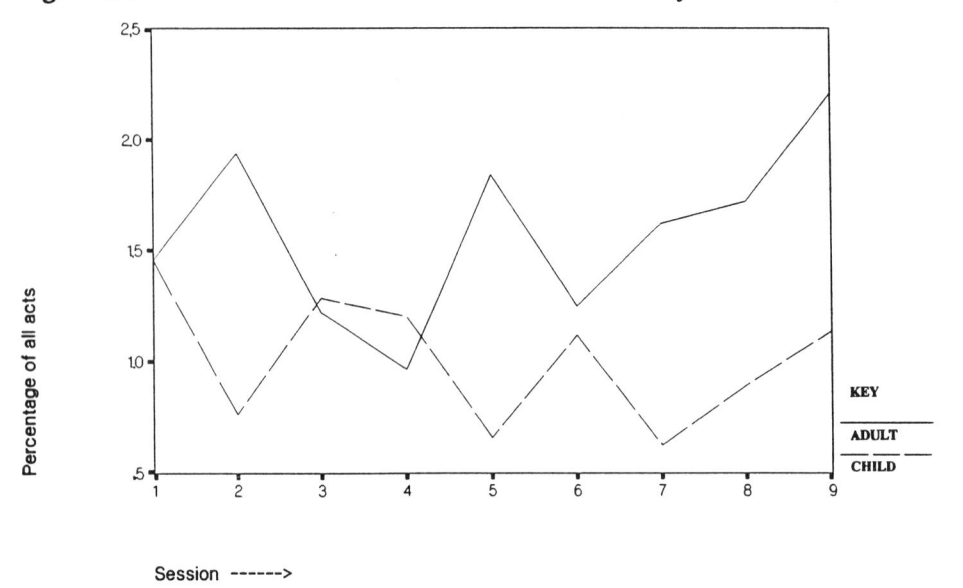

Figure 6.6b. Self-related imitations of adult in SLI dyads

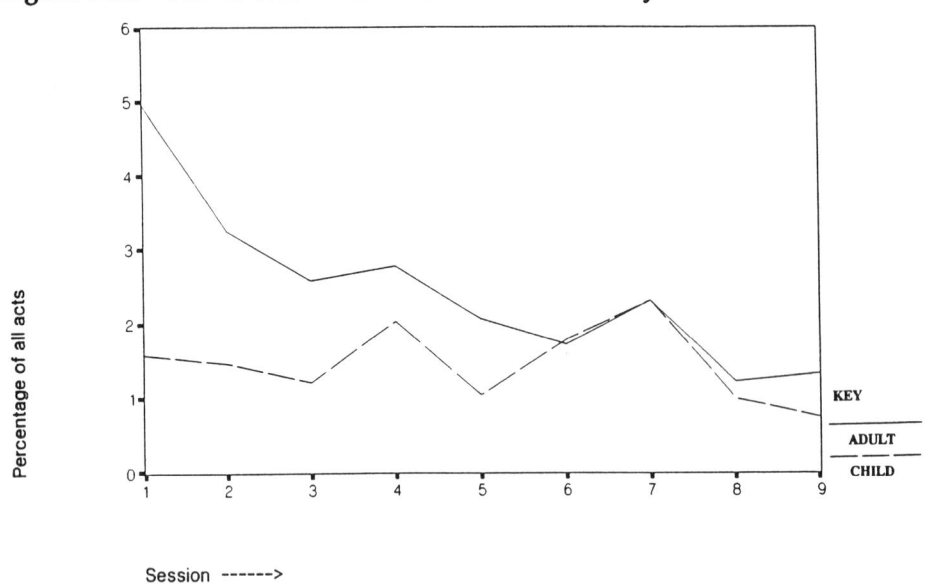

Session Effect
The Tables F2 (Appendix F) and 6.2. indicate that there are also no significant 'Session Effect' for partner-related imitations in both groups of children. Partner-related imitations used by the SLI children tend to decrease (not significantly) from session 1 to 9 (Figure 6.7a,b).

Figure 6.7a. Partner-related imitations child in NLA dyads

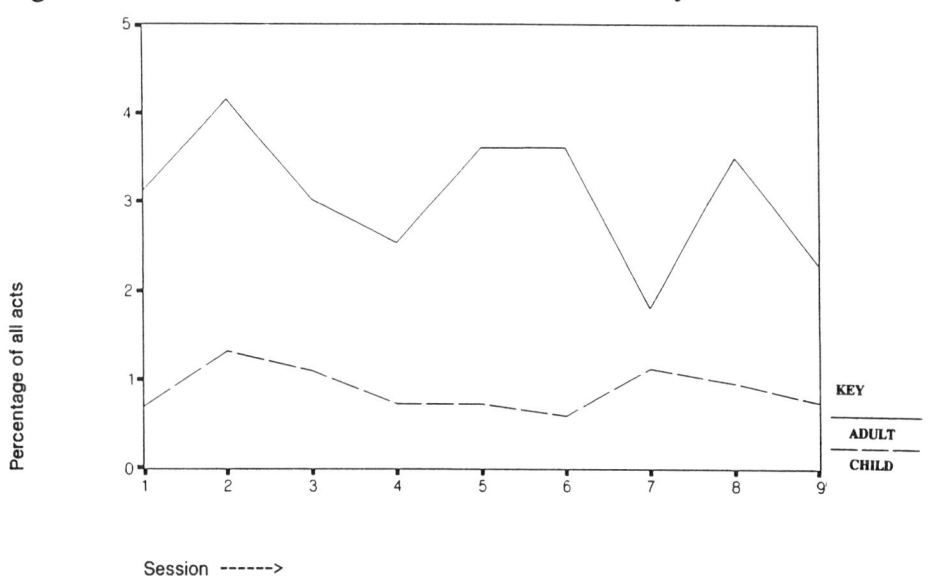

Figure 6.7b. Partner-related imitations child in SLI dyads

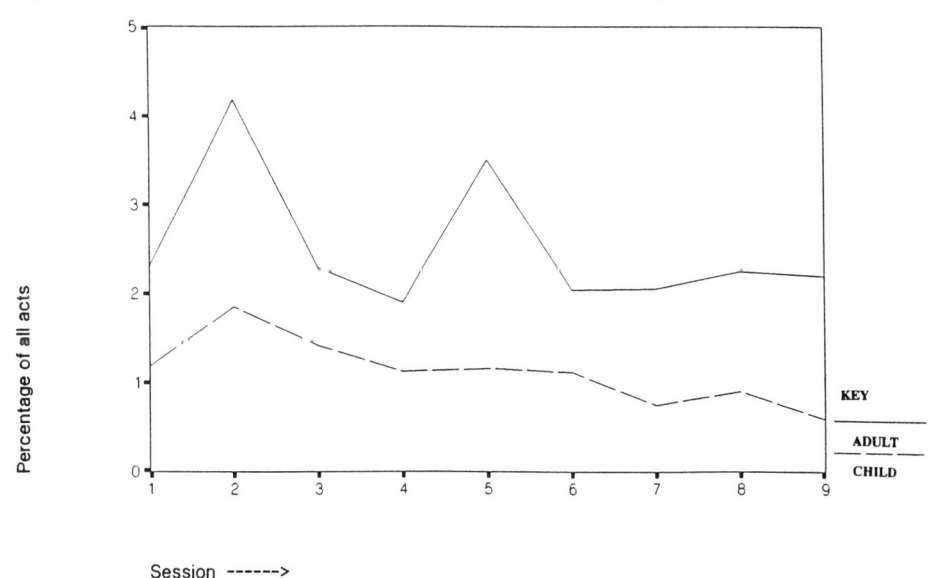

6.2.8. Nonverbal Initiations
Between Groups Effect
The Tables F2 (Appendix F) and 6.2. indicate a significant 'Between Groups Effect' for nonverbal initiations of the NLA and the SLI children. The SLI children tend to use more nonverbal initiations than their caregivers and use significantly more nonverbal initiations than the NLA children. The NLA children tend to use more verbal initiations than their caregivers and use significantly more verbal initiations than the SLI children. The caregivers of the SLI children use more verbal initiations than their children. The adults in the SLI dyads use significantly more initiations than the caregivers of the NLA children (see also section 6.2.4.).

Session Effect
The Tables F2 (Appendix F) and 6.2. indicate a significant 'Session Effect' for the nonverbal initiations of the SLI children. That 'Session Effect' is mainly produced by an increase in nonverbal initiations of the SLI children from session 1 to 7 and a decrease from session 7 to 9 (Figure 6.8b.).

Figure 6.8a. Nonverbal initiations in NLA dyads

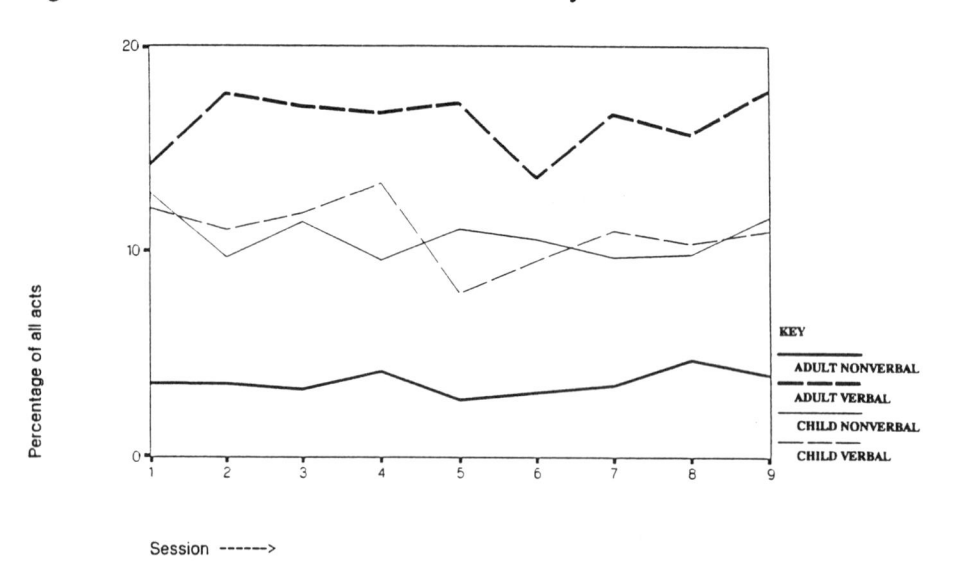

Figure 6.8b. Nonverbal initiations in SLI dyads

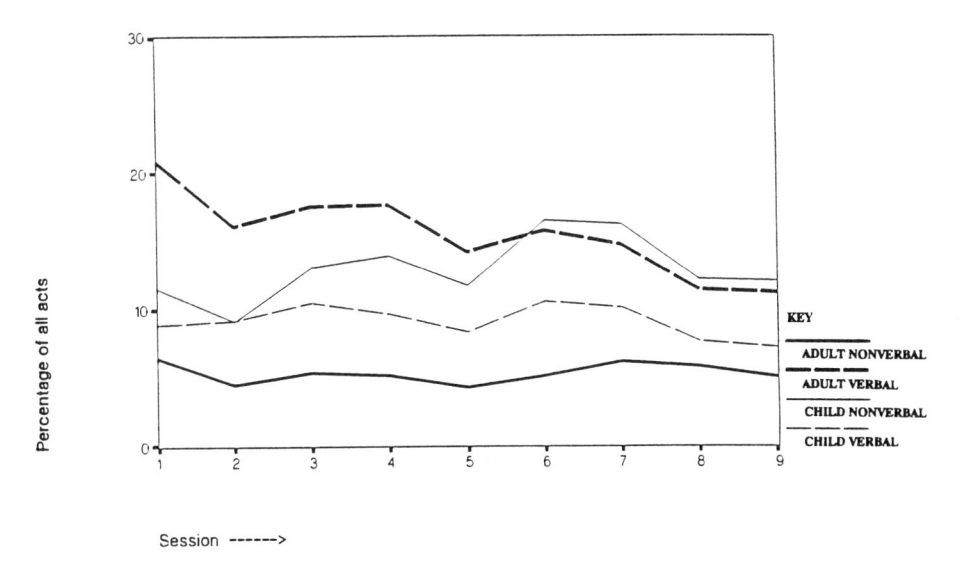

6.2.9. Summary

In the preceding section the results of the IACV categories related to the hypotheses on 'Turn-taking' were presented. In this summary section an overview of the hypotheses tested is given.

Hypotheses: TURN-TAKING Compared with NLA children of the same age and their caregivers:	Test Result
1a. SLI children use more backchannelling	confirmed
1b. SLI children use more ellipsis	rejected
2. caregivers of SLI children have longer turns	rejected
3a. caregivers of SLI children initiate themes more often	rejected
3b. caregivers of SLI children reinitiate themes more often	confirmed
4. caregivers of SLI children use more self-repetitions	confirmed
5. SLI children use more imitations of preceding adult acts	rejected
6. initiative behaviour of SLI children is mostly nonverbal	confirmed

The interpretation of the results presented in the previous sections will be given in chapter 7 (section 7.2.).

6.3. Child Adjusted Register (CAR)

TABLE 6.3. Results of MANOVA's for Session and Between Groups Effect: 'Child Adjusted Register (CAR)'

		SESSION EFFECT		GROUPS EFFECT
Hypotheses CAR		SLI Sign.of F (df 88/df 8)	NLA Sign. of F (df 40/df 8)	Sign. of F (df 16/df 1)
MLU	Adult	0.117	0.665	0.115
	Child	0.000***	0.032*	0.001***
Imitations, self-related Adult		0.000***	0.612	0.002**
Requests for Information Adult		0.001***	0.243	0.227

* $p \leq 0.05$
** $p \leq 0.01$
*** $p \leq 0.001$

6.3.1. Mean Length of Utterance (MLU)
Between Groups Effect

The Table F3 (Appendix F) and 6.3. indicate that the caregivers of the SLI and the NLA children do not differ significantly in their MLU, although the caregivers of the NLA children tend to use utterances with a higher MLU in the first five sessions (see also Figure 6.9.). The average MLU of the NLA and the SLI children show a significant 'Between Groups Effect'. The average MLU of the NLA children is much higher than the average MLU of the SLI children at the first five sessions.

As Figure 6.9. shows, the differences in the average MLU between the SLI and the NLA children decrease from session 6 to 9, due to a general increase in MLU of the SLI children.

Figure 6.9. Mean length of utterance (MLU) in NLA and SLI dyads

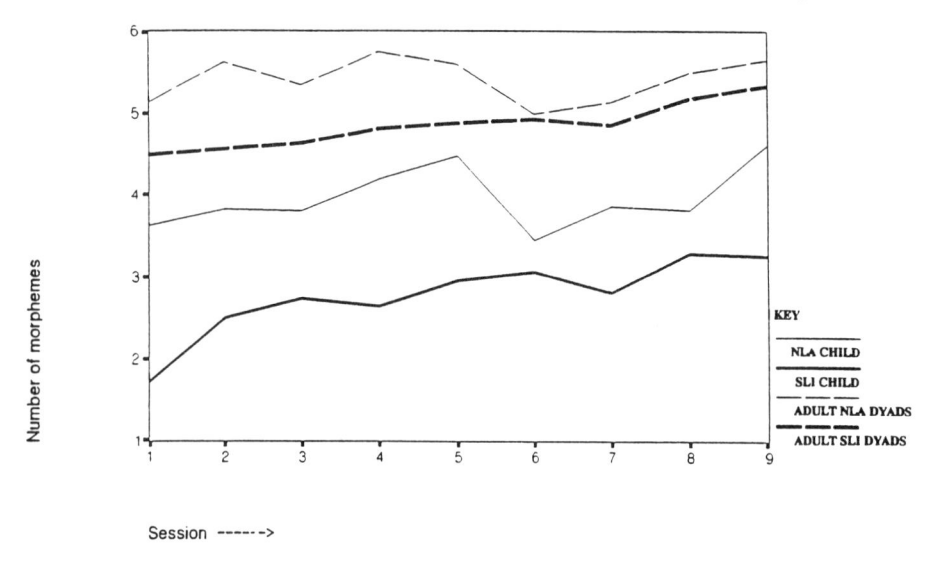

Session Effect
The average MLU of the NLA and the SLI children show significant 'Session Effect'. Both the NLA and the SLI children show significant 'Session Effect', indicating an increase in the average MLU for the NLA children and for the SLI children. The difference in the average MLU between the caregivers and the children in the SLI dyads remains larger than the difference in the average MLU between the caregivers and the children in the NLA dyads from session 1 to 9 (see also the MLU in the Tables 5.1, 5.2., 5.3. and 5.4. in section 5.2.2.)

6.3.2. Self-related Imitations Adult
Between Groups Effect
The Tables F3 (Appendix F) and 6.3. and the Figures 6.6a,b (in section 6.2.6.) indicate a significant 'Between Groups Effect' for caregivers' self-related imitations. The caregivers of SLI children use significantly more self-related imitations than the caregivers of the NLA children, primarily during the first session.

Session Effect
The Tables F3 (Appendix F) and 6.3. and the Figures 6.6a,b (in section 6.2.6.) indicate a significant 'Session Effect' for caregivers' self-related imitations in the SLI dyads. The number of self-related imitations used by the caregivers in the SLI dyads decreases after session 1, but still remains higher than number of self-related imitations of the caregivers in the NLA dyads.

6.3.3. Requests for Information Adult

Requests for information were coded in the group of 'Representational Communication Functions' (for definitions, see Appendix C).

Between Groups Effect

The caregivers in the SLI and in the NLA dyads do not differ significantly in the number of information requests used.

Session Effect

The Tables F3 (Appendix F) and 6.3. indicate a significant 'Session Effect' for information requests from the caregivers in the SLI dyads. The number of adult information requests used in the SLI dyads decreases from session 4 to 9 (with intermediate peaks at session 6 and 7, due to individual effects in DN 3, 5, 6 and 12) and tends to increase (not significantly) from session 6 to 9 (with an intermediate dip at session 8, due to individual effects in DN 15) in the NLA dyads.

6.3.4. Summary

In the preceding section the results of the IACV categories relating to the hypotheses on 'Child Adjusted Register (CAR)' were presented. In this summary section an overview of the hypotheses tested is given.

Hypotheses: CHILD ADJUSTED REGISTER (CAR) Compared with NLA children of the same age, SLI children receive:	Test Result
1. a language input, characterised by a smaller MLU	rejected
2. a language input consisting of more (adult) self-imitations	confirmed
3. a language input consisting of more (adult) requests for information.	rejected

The interpretation of the results presented in the previous sections will be given in chapter 7 (section 7.3.)

6.4. Topic and Theme Management

TABLE 6.4. Results of MANOVA's for Session and Between Groups Effect: 'Topic and Theme Management'

		SESSION EFFECT		GROUPS EFFECT
Hypotheses Topic/Theme		SLI Sign.of F (df 88/df 8)	NLA Sign. of F (df 40/df 8)	Sign. of F (df 16/df 1)
Introduction New Theme	Adult	0.004**	0.824	0.468
	Child	0.000***	0.509	0.833
Mean Length of Subsequence	Adult	0.000***	0.830	0.060
	Child	0.000***	0.941	0.398
Imitations, self-related Adult		0.000***	0.612	0.002**
Imitations, partner-related Child		0.206	0.720	0.720

* $p \leq 0.05$
** $p \leq 0.01$
*** $p \leq 0.001$

6.4.1. Introduction New Theme
Initiations are classified according to three possible subcategories: introduction of a new theme, elaboration of a theme and reintroduction (or reinitiation) of a theme (definitions and examples were given in section 4.3.1.). Data concerning the elaborations and reinitiations of themes were already presented in section 6.2.5. (Turn-taking). In this section the group of initiations coded as 'Introduction of New Theme' is discussed.

Between Groups Effect
Caregivers and children in the SLI and in the NLA dyads do not differ significantly in the number of 'Introduction of New Themes' used.

Session Effect

Table F4 (Appendix F) and 6.4. and the Figures 6.5a,b (Adult Reinitiations and Elaborations) indicate significant 'Session Effect' for the use of 'Introduction of New Themes' by the caregivers and the children in the SLI dyads.

Figure 6.10a. Initiations, elaborations, reinitiations of NLA child

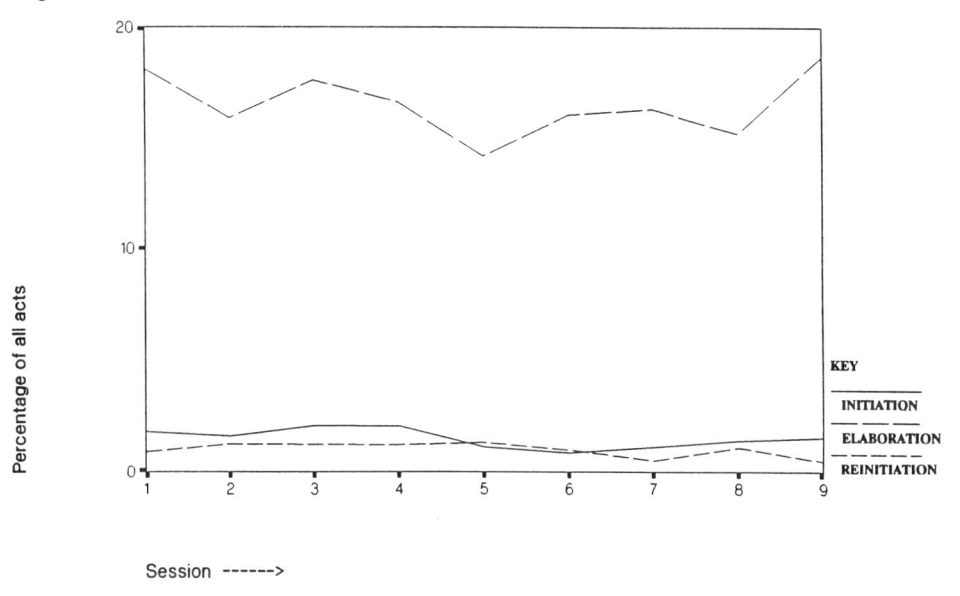

Figure 6.10b. Initiations, elaborations, reinitiations of SLI child

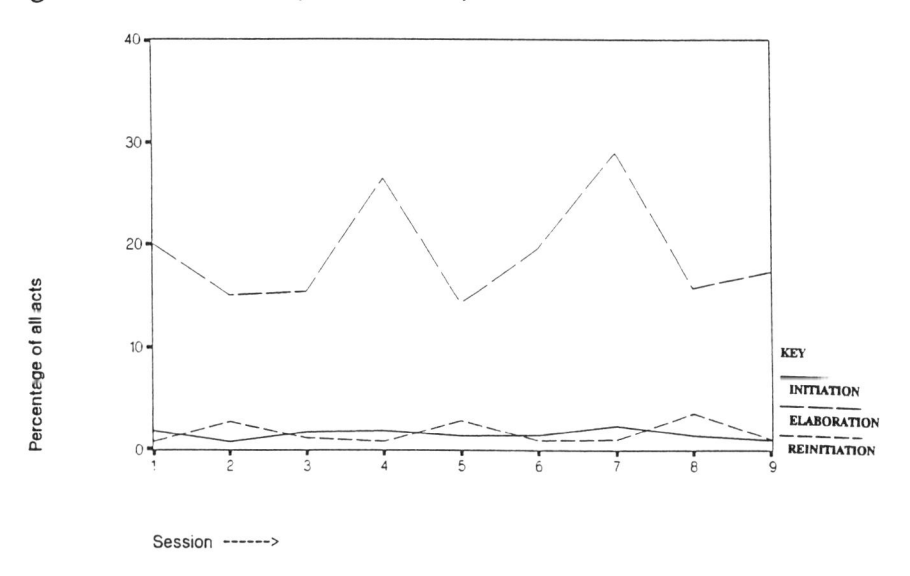

6.4.2. Mean Length of Subsequence (MLS)
Between Groups Effect
There are no significant differences in the 'Mean Length of Subsequence (MLS)' of the caregivers and the children in the SLI and in the NLA dyads. Generally the subsequences initiated by the SLI children tend to be longer than the subsequences initiated by their caregivers (Figures 6.11a,b).

Session Effect
The Tables F4 (Appendix F) and 6.4. indicate that there are significant 'Session Effect' in the MLS for the caregivers and for the children in the SLI dyads. Both the 'Session Effect' for the MLS of the caregivers and for the MLS of the children in the SLI dyads are mainly the result of an increase in the MLS from session 7 to 9.

Figure 6.11a. Mean length of subsequence in NLA dyads

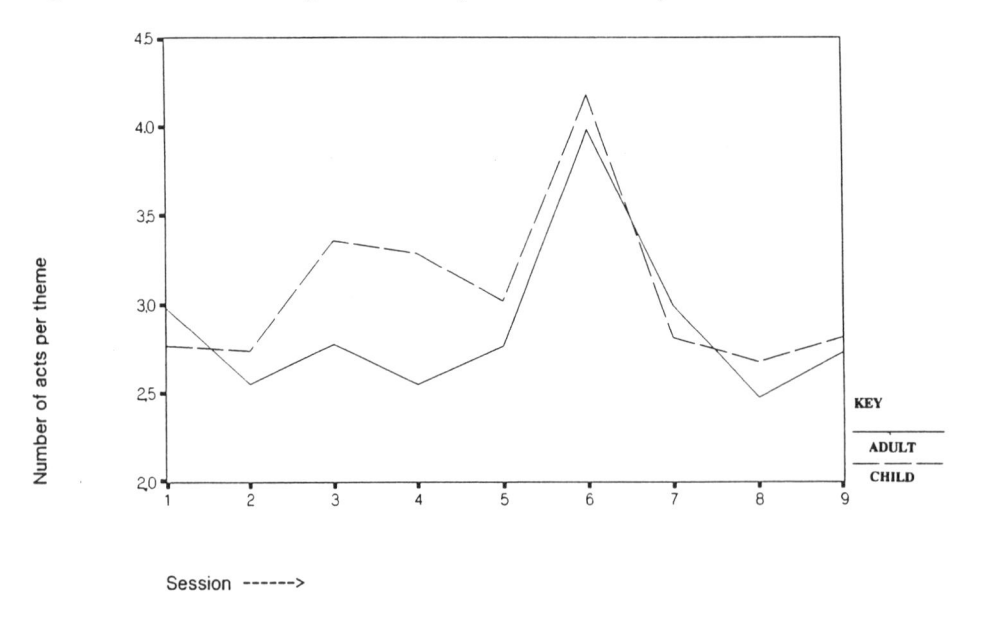

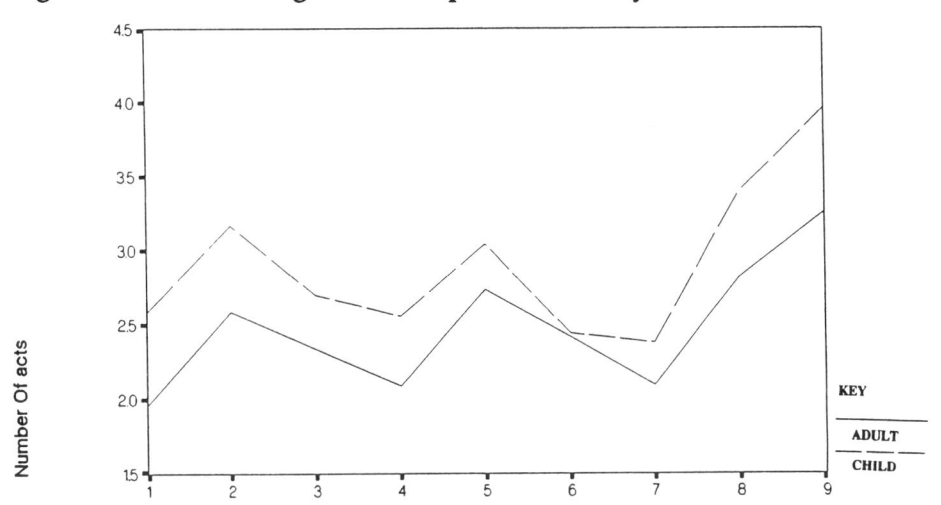

Figure 6.11b. Mean length of subsequence in SLI dyads

6.4.3. *Partner-related Imitations Child Between Groups Effect*
As already indicated in section 6.2.7., there is no significant 'Between Groups Effect' for partner-related imitations of the NLA and of the SLI children.

Session Effect
As already indicated in section 6.2.7., there is no significant 'Session Effect' for partner-related imitations used by the NLA and by the SLI children. Partner-related imitations used by the SLI children tend to decrease (not significantly) from session 1 to 9.

6.4.4. *Self-related Imitations Adult*
Between Groups Effect
As already indicated in section 6.2.6, there is a significant 'Between Groups Effect' for caregivers' self-related imitations in the NLA and in the SLI dyads. The caregivers of the SLI children use significantly more self related imitations than the caregivers of the NLA children.

Session Effect
As already indicated in section 6.2.6, there is a significant 'Session Effect' for caregivers' self-related imitations in the SLI dyads. That 'Session Effect' is mainly a result of a significant decrease in the caregivers' self-related imitations in the SLI dyads.

6.4.5. Summary

In the preceding sections the results of the IACV categories relating to the hypotheses on 'Topic and Theme Management' were presented. In this summary section an overview of the hypotheses tested is given.

Hypotheses: TOPIC AND THEME MANAGEMENT Compared with NLA children of the same age, the communicative interaction between SLI children and their caregivers is characterised by:	Test Result
1. more child initiations of 'introductions new themes'	rejected
2. shorter themes in general	rejected
3. shorter themes, initiated by caregivers of SLI children	rejected
4. more imitations of adult acts by the child	rejected
5. more imitations of self-related acts by the adult	confirmed

The results presented in the previous sections are discussed in chapter 7 (section 7.4.).

6.5. Communication Functions

6.5.1. Requests for Information Adult and Child
Between Groups Effect

The Tables 6.5. and F3 and F5 (Appendix F) indicate that there is no significant difference in the frequencies of information requests used by the caregivers in the SLI and by the caregivers in the NLA dyads. There is also no significant 'Between Groups Effect' for the NLA and the SLI children in the number of information requests used.

Session Effect

The Tables 6.5. and F3 and F5 (Appendix F) indicate a significant 'Session Effect' for information requests used by the caregivers in the SLI dyads. The number of adult information requests used in the SLI dyads decreases (significantly) from session 4 to 9 and tends to increase (not significantly) from ses-

TABLE 6.5. Results of MANOVA's for Session and Between Groups Effect: Communication Functions

		SESSION EFFECT		GROUPS EFFECT
Hypotheses Commun. Functions		SLI Sign. of F (df 88/df 8)	NLA Sign. of F (df 40/df 8)	Sign. of F (df 16/df 1)
Information Requests				
	Adult	0.001***	0.665	0.115
	Child	0.743	0.493	0.833
Controls	Child	0.049*	0.259	0.253
	Adult	0.187	0.062	0.488
Expressives	Child	0.014*	0.182	0.096
	Adult	0.123	0.526	0.654
Representationals				
	Child	0.008**	0.049*	0.523
	Adult	0.053	0.112	0.419
Tutorials	Child	0.361	0.325	0.223
	Adult	0.518	0.627	0.177
Procedurals	Child	0.000***	0.247	0.615
	Adult	0.000***	0.139	0.665

* $p \leq 0.05$
** $p \leq 0.01$
*** $p \leq 0.001$

'Speech for Self' and 'Socials' showed no variance and are ommitted from this table.

sion 6 to 9 in the NLA dyads (see also section 6.3.3.). There is also no significant 'Session Effect' in the number of information requests used by the NLA and by the SLI children. The number of information requests of the SLI children tends to increase (not significantly) from session 1 to 4 and tends to decrease (not significantly) from session 5 to 9.

6.5.2. *Communication Functions Adult and Child*
Between Groups Effect
According to the Tables F5 (Appendix F) and 6.5. the SLI and the NLA children show no significant 'Between Groups Effect' for all communication functions.

Session Effect

According to the Tables F5 (Appendix F) and 6.5. the SLI children show a significant 'Session Effect' for controls (mainly due to a general decrease in controls from session 1 to 9), expressives (mainly due to a sudden increase in expressives at session 7 and 8), representationals (mainly due to a general increase in representationals from session 1 to 9) and procedurals (mainly due to variations in the frequencies of procedurals per session).

The NLA children show a significant 'Session Effect' for representationals only (mainly due to an increase in the use of representationals from session 5 to 6 and a decrease in the use of representationals from session 6 to 8).

There is also a significant 'Session Effect' for the caregivers of the SLI children in their use of representationals and a significant 'Session Effect' for procedurals, mainly as the result of a general increase in representationals and peaks for procedurals at the sessions 2 (mainly due to individual effects in DN 2 and DN 12), 5 (mainly due to individual effects in DN 9) and 8 (mainly due to individual effects in DN 2 and DN 6). The number of controls used by the caregivers in the SLI dyads tends to decrease (not significantly) from session 1 to 9.

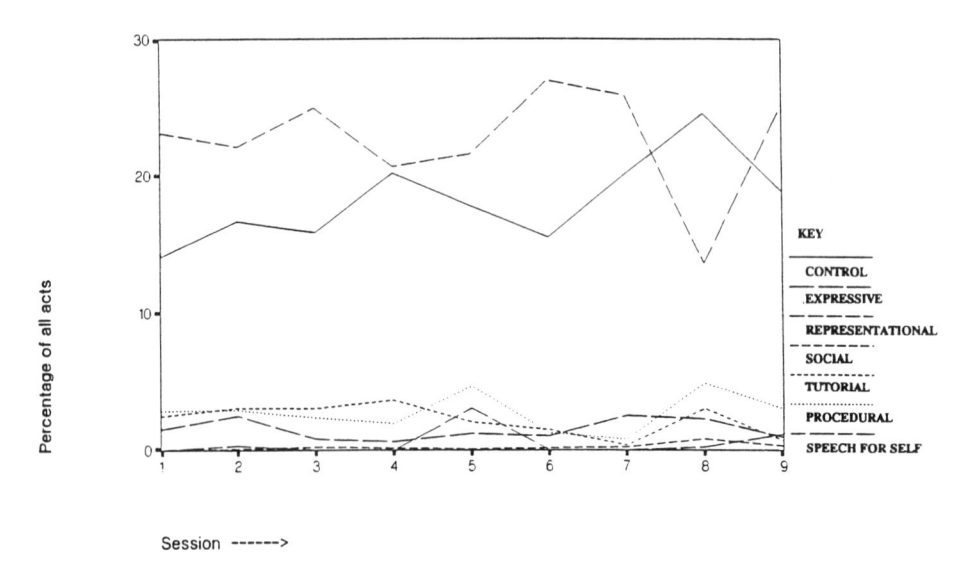

Figure 6.12a. Communication functions of subsequences adult in NLA dyads

Figure 6.12b. Communication functions of subsequences adult in SLI dyads

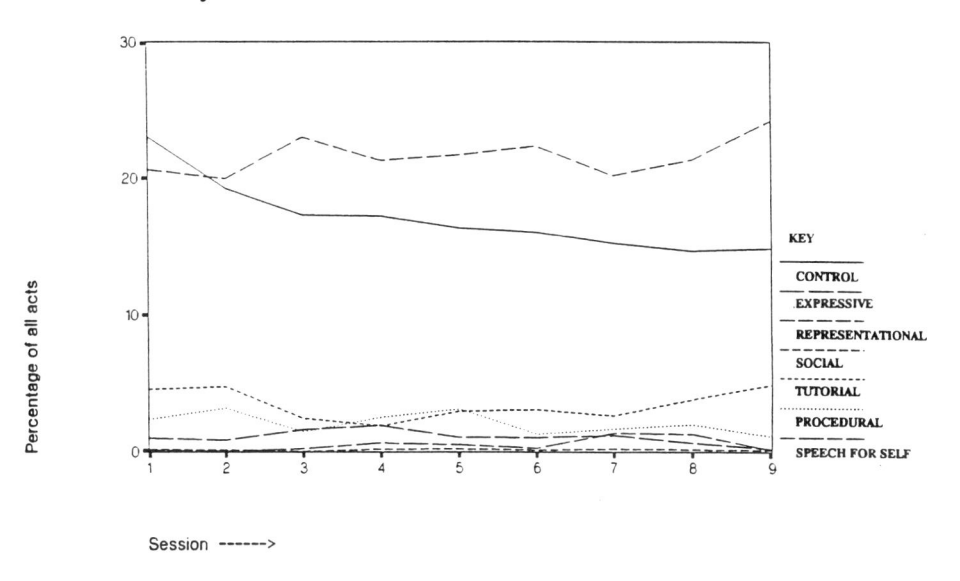

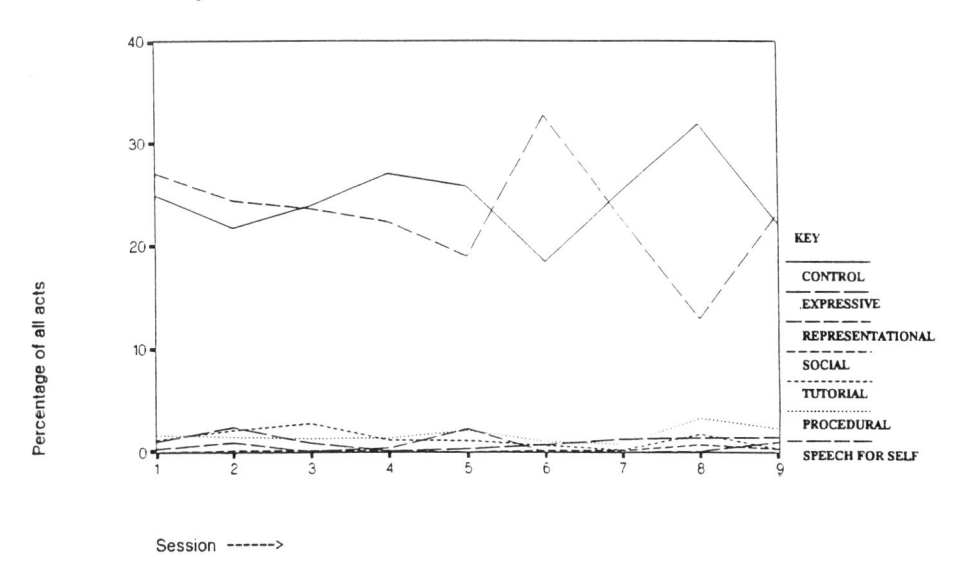

Figure 6.12c. Communication functions of subsequences child in NLA dyads

Figure 6.12d. Communication functions of subsequences child in SLI dyads

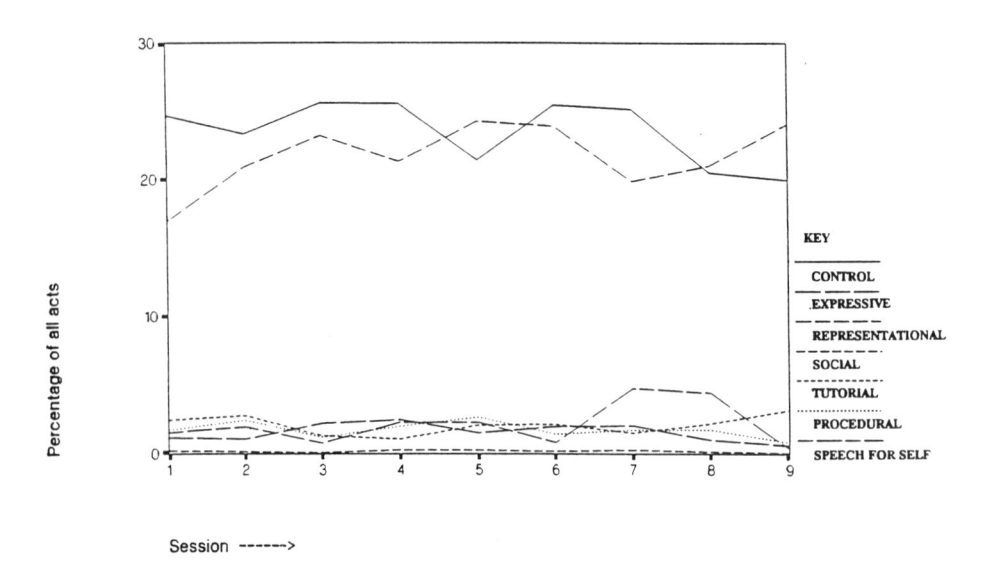

6.5.3. Summary

In the preceding section the results of the IACV categories relating to the hypotheses on 'Communication Functions' were presented. In this summary section an overview of the hypotheses tested is given.

Hypotheses: COMMUNICATION FUNCTIONS NLA children compared with SLI children of the same age	Test Result
1. use fewer requests for information	rejected
2a. receive a language input with more controls	rejected
2b. receive a language input with more tutorials.	rejected

The results presented in the previous sections are discussed in chapter 7 (section 7.5.).

6.6. Communication Breakdowns and Repairs

TABLE 6.7. Results of MANOVA's for Session and Between Groups Effect: Communication Breakdowns and Repairs

	SESSION EFFECT		GROUPS EFFECT
Hypotheses Breakdowns/Repairs	SLI Sign.of F (df 88/df 8)	NLA Sign. of F (df 40/df 8)	Sign. of F (df 16/df 1)
Unfinished verbal acts Child	0.336	0.090	0.627
Parallel Talk	—	0.013*	0.541
Clarification Requests Adult	0.000***	0.213	0.012*
Corrections Adult	0.000***	0.745	0.011*
Consecutive Initiations	0.000***	0.966	0.263
Faulty Responses Adult	0.913	0.217	0.023*
Faulty Responses Child	—	0.037*	0.272
Faulty Initiations Adult	0.000***	0.016*	0.012*
Faulty Initiations Child	0.000***	0.216	0.285

* $p \leq 0.05$
** $p \leq 0.01$
*** $p \leq 0.001$

6.6.1. Unfinished Verbal Acts
Between Groups Effect

The Tables 6.7. and F6 (Appendix F) indicate that there is no significant 'Between Groups Effect' in the use of unfinished verbal acts. The SLI and the NLA children do not differ significantly as a group in the number of unfinished verbal acts used.

Session Effect
The Tables 6.7. and F6 (Appendix F) indicate that there is no significant 'Session Effect' in the number of unfinished verbal acts used by the SLI and by the NLA children. The number of unfinished acts used by the NLA children tends to decrease (not significantly) from session 1 to 9. On the other hand, the number of unfinished verbal acts used by the SLI children tends to increase (not significantly) from session 1 to 9.

6.6.2. Parallel Talk
Between Groups Effect
The Tables F6 (Appendix F) and 6.7. indicate that the SLI and the NLA dyads do not differ significantly in the number of parallel talk used.

Session Effect
The Tables F6 (Appendix F) and 6.7. indicate a significant 'Session-Effect' for parallel talk in the NLA dyads, mainly due to individual effects in DN 13 at session 9.

6.6.3. Adult Clarification Requests
Between Groups Effect
The Tables F6 (Appendix F) and 6.7. indicate a significant 'Between Groups Effect' for the number of adult clarification requests used in the SLI dyads. The 'Between Groups Effect' is mainly a result of a general higher number of clarification requests by the caregivers in the SLI dyads.

Session Effect
The Tables F6 (Appendix F) and 6.7. indicate a significant 'Session Effect' for the number of adult clarification requests used in the SLI dyads. The 'Session Effect' in the SLI dyads is mainly a result of a general decrease in the use of adult clarification requests from session 7 to session 9 (especially in DN 5 and DN 7).

Figure 6.13a. Clarifications requests in NLA dyads

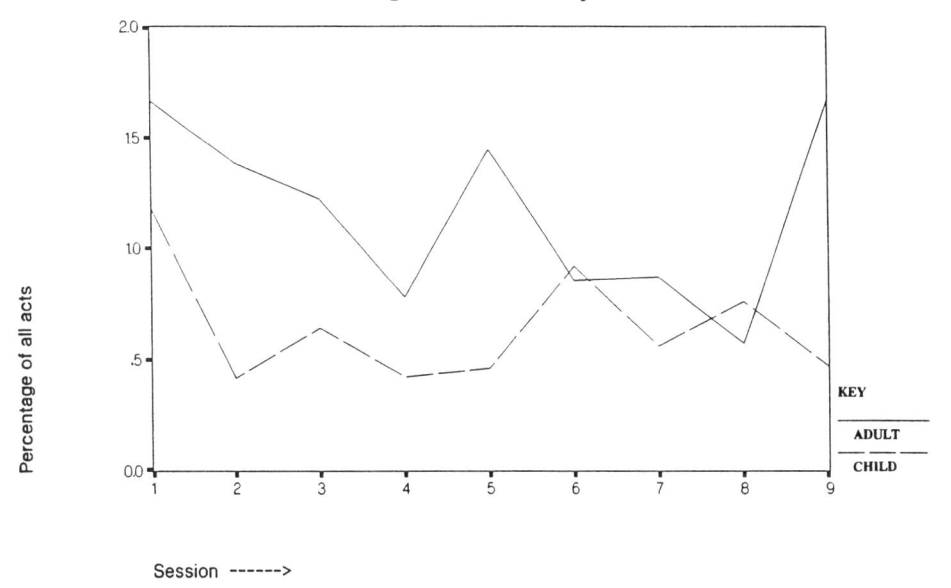

Figure 6.13b. Clarifications requests in SLI dyads

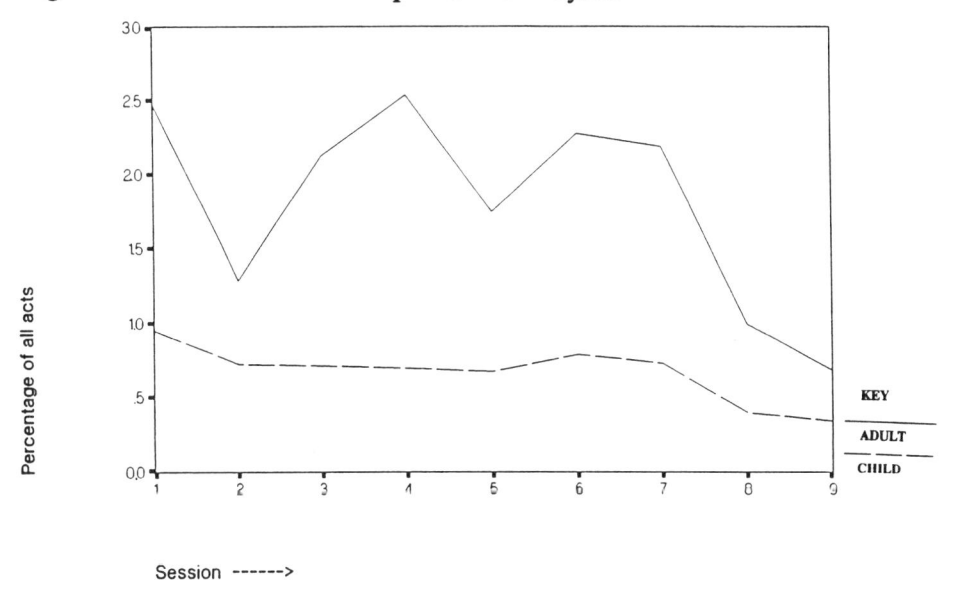

6.6.4. Adult Corrections of Previous Act
Between Groups Effect

The Tables F6 (Appendix F) and 6.7. indicate a significant 'Between Groups Effect' for the number of corrections used by the caregivers in the SLI dyads. The 'Between Groups Effect' is mainly a result of a general higher number of caregivers' corrections in the SLI dyads.

Session Effect
The Tables F6 (Appendix F) and 6.7. indicate a significant 'Session Effect' for the number of corrections used by the caregivers in the SLI dyads. The SLI 'Session Effect' is mainly a result of a general decrease in the adults' corrections from session 1 to 9 (with an intermediate peak at session 7, due to individual effects in DN 3).

6.6.5. Consecutive Initiations
Between Groups Effect
The Tables F6 (Appendix F) and 6.7. indicate no significant differences in the number of consecutive initiations used in the conversations of caregivers and children in the SLI and in the NLA dyads. Generally consecutive initiations appear more often (not significantly) in the SLI dyads than in the NLA dyads.

Session Effect
The Tables F6 (Appendix F) and 6.7. indicate a significant 'Session Effect' for consecutive initiations in the SLI dyads, mainly caused by three dips in frequencies of consecutive initiations at session 2 (primarily in DN 2, DN 4, DN 6, DN 8, DN 9 and DN 11), 5 (primarily in DN 3, DN 6, DN 8, DN 9, DN 12) and a general decrease from session 8 to 9.

6.6.6. Faulty Responses
Between Groups Effect
The Tables F6 (Appendix F) and 6.7. indicate a significant 'Between Groups Effect' for the faulty adult responses. This 'Between Groups Effect' is mainly a result of a higher incidence of faulty adult responses in the SLI dyads. Generally, the number of faulty responses from the caregivers and the children is very small, this makes a meaningful discussion about the importance of the observed significant effects very difficult.

Session Effect
The Tables F6 (Appendix F) and 6.7. indicate a significant 'Session Effect' for faulty child responses in the NLA dyads, mainly due to the variations in the frequencies of faulty child responses per session.

6.6.7. Faulty Initiations
Between Groups Effect
The Tables F6 (Appendix F) and 6.7. indicate a significant 'Between Groups Effect' for the caregivers' faulty initiations, indicating that the caregivers in the SLI dyads use faulty initiations more often than the caregivers in the NLA dyads.

Session Effect
The Tables F6 (Appendix F) and 6.7. indicate significant 'Session Effect' for faulty initiations by the caregivers in the SLI and in the NLA dyads and for the SLI children, mainly due to the variations in frequencies per session.

Figure 6.14a. Faulty initiations in NLA dyads

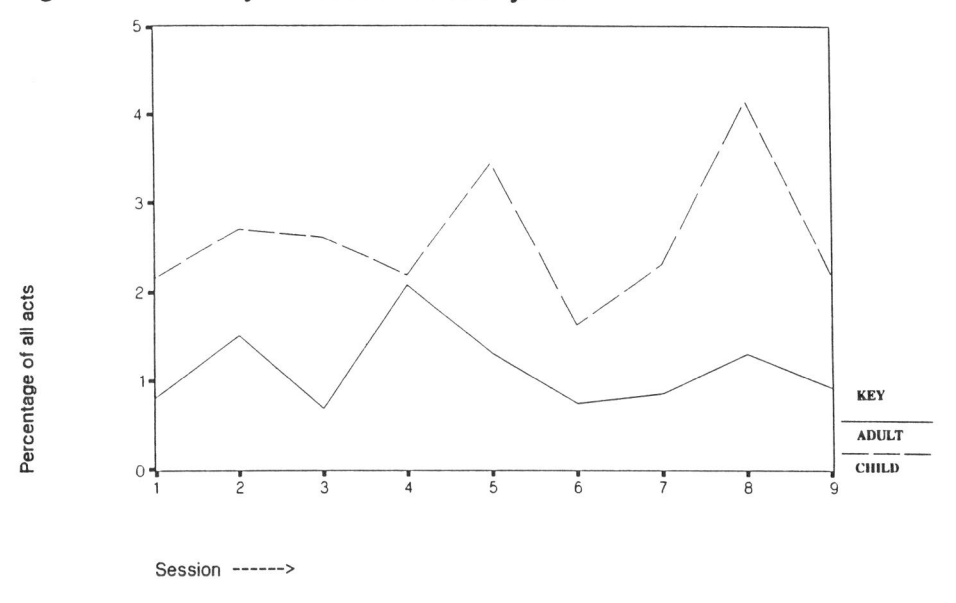

Figure 6.14b. Faulty initiations in SLI dyads

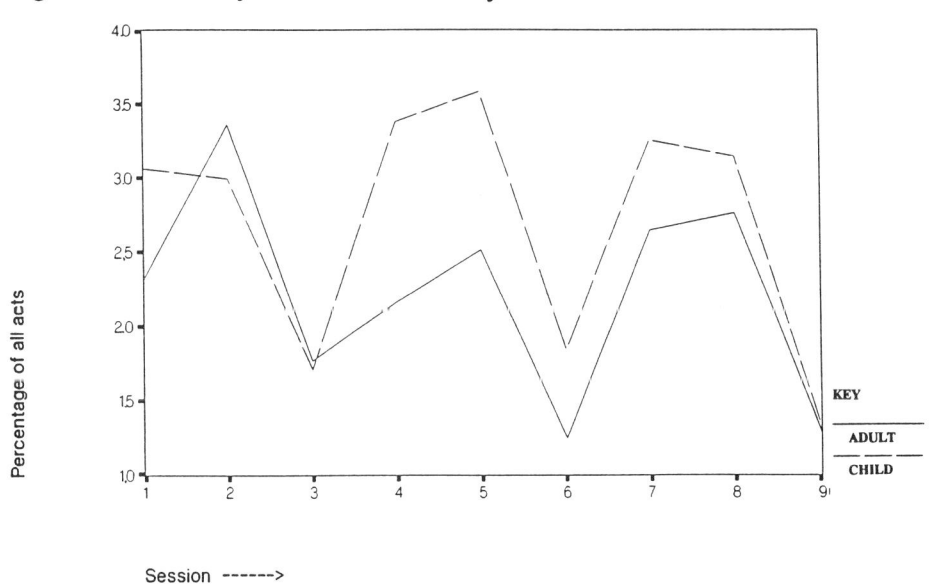

6.6.8. Summary

In the preceding section the results of the IACV categories relating to the hypotheses on 'Communication Breakdowns and Repairs' were presented. In this summary section an overview of the hypotheses tested is given.

Hypotheses: COMMUNICATION BREAKDOWNS AND REPAIRS Compared with NLA child-caregiver interaction, SLI child-caregiver conversations are characterised by:	Test Result
1. more unfinished verbal acts of SLI children	rejected
2. a high number of parallel talk and verbal-nonverbal overlaps of both partners, with different communicative functions	rejected
3. more clarification requests and corrections used by adults	confirmed
4. more consecutive initiations	rejected
5a. more faulty responses from child to preceding acts of adult	rejected
5b. more faulty responses from adult to preceding acts of child	confirmed
6a. more faulty initiations of adult	confirmed
6b. more faulty initiations of child	rejected

In the next chapter the results presented in the previous sections are discussed in the light of the literature reviewed in chapter 3.

Chapter 7 has an arrangement of sections similar to chapter 6.

CHAPTER 7

Discussion

In this chapter the results presented in the previous chapter are discussed by referring to the literature on discourse coherence in adult-child conversations reviewed in chapter 3. The discussion follows the order of the hypotheses described in section 4.5. The hypotheses are restated at the beginning of each section (starting at section 7.2.).

7.1. Total Number of Acts used by Adult and Child

The current investigation reveals a significant 'Between Groups Effect' for the total number of children's acts and the total number of children's nonverbal acts (Table 6.1.). The number of nonverbal acts used by the SLI children increases from session 1 to 7 and decreases from session 7 to 9 (not significantly) (see Figure 6.1b). The number of nonverbal acts used by the NLA children continues to be stable from session 1 to 9 (see Figure 6.1a).

The SLI children use significantly more nonverbal acts than the NLA children. The nonverbal acts used by the SLI children can be interpreted as a strategy to compensate their limited verbal and linguistic abilities (Bates et al.,

1975). The SLI children's involvement in communicative interaction increases from session 1 to 9, mainly due to the steep increase in nonverbal acts and a slight increase in their verbal acts (Table F1, Figure 6.1b.). The active involvement of the caregivers in the SLI dyads decreases from session 1 to 9, mainly indicated by a (non significant) decrease in the caregivers' verbal acts and no (significant) changes in nonverbal performance. The SLI children seem to be willing to participate actively in communicative interaction, but do not have full access to verbal and linguistic skills. This seems to explain the increase in nonverbal behaviour and its continuous importance in the SLI dyads compared with the NLA dyads.

With an increase in the average number of acts performed by the children it is expected that there is also an increase in the average number of acts addressed to the children (Schachter, 1979; Wells, 1985). In contrast to this expectation the data of the study at hand show a decrease in caregivers' acts in the SLI dyads. This seems to indicate a disbalanced mutual involvement in discourse activities of adult and child in the SLI dyads. In the NLA dyads the mutual discourse involvement of both partners and the average number of verbal and nonverbal acts used by the partners remain almost equal and stable from session 1 to session 9. Similar to the SLI children the majority of the acts used by the NLA children is nonverbal. However there is no significant difference between the mean number of verbal and nonverbal acts used by the NLA children. The relative stable average number of verbal acts used by the NLA children from session 1 to 9 (from the NLA mean age 2;11 to 4;06) is also reported in other studies (Wells, 1985, 112). In the NLA dyads more than 50% of the average number of acts remains verbal (Figure 6.1a.). In the SLI dyads the average number of verbal acts decreases from about 60% in the first three sessions to 40% from session 5 to session 9, mainly caused by the decrease in the average number of verbal acts of the caregivers (Figure 6.1b.). The conversations in the SLI dyads are determined by nonverbal acts of the children. The conversations in the NLA dyads primarily consist of verbal acts of the caregiver.

The SLI dyads offered significantly fewer instances for the caregivers to react verbally, which partly explains the decrease in the number of caregivers' verbal acts used from session 1 to 9. When comparing the percentages of verbal acts used by the caregivers in the SLI and in the NLA dyads during the first two sessions the differences are small. The differences in the numbers of verbal acts used become more obvious in session 2. Both the NLA and the SLI children are in their early stages of language development at the beginning of the current study. In these stages of early language acquisition caregivers guide their children's verbal language performance in conversations by using so-called 'proto-conversations' (Bateson, 1975; McTear, 1985).

Caregivers often synchronise their turns with what the child is doing or likely to do in such a way that the child's actions are seen as responses. The similarity in the number of caregivers' verbal acts used in the SLI and in the NLA dyads during the first sessions can be attributed to the early stages of language development of the NLA and of the SLI children. The SLI children display difficulties in verbal performance and linguistic development (as discussed in chapter 5, section 5.2.2.) and they seem to compensate this inability by using more nonverbal acts (Friel-Patti & Conti-Ramsden, 1984; McTear, 1985).

7.2. Turn-taking

7.2.1. Backchannelling
Hypothesis:
SLI children use more backchannelling than NLA children (confirmed)
The hypothesis stating that the SLI children use more backchannels than the NLA children is confirmed by the data of this study (Table 6.2.). Table F2 and the Figures 6.2a,b. indicate a higher average number of backchannels used by the SLI children. Other studies showed similar effects (Watson, 1977, Fey et al, 1981, Blau, 1986). In these studies it was suggested that SLI children rely more heavily on backchannelling as a means of avoiding a turn in spontaneous conversations with their caregivers. This may also be true in the SLI dyads investigated in the current study. However, a high (average) number of backchannels does not necessarily mean conversational 'passivity' or 'incompetence'. In the study at issue turn-taking also incorporates nonverbal initiations and responses. The problem with the studies on backchannelling mentioned above (except the study reported by Blau, 1986) is their primary focus on verbal acts in general and verbal turn-taking behaviour in particular.

Contrary to the often reported results in the studies of discourse involvement, turn-taking and backchannelling in conversations between caregivers and their SLI children (Watson, 1977; Fey et al, 1981), the study at issue indicates that the SLI children in fact are active (nonverbal) discourse participants (see also the sections 7.2.7. and 7.2.8.). The set-up of the current study differs from the above mentioned investigations mainly in the way of analysing the discourse sequentially with equal consideration of nonverbal and verbal acts of both partners. In general the study indicates that the SLI children, compared with the NLA children, rely more heavily on backchannels in order to cohere with their caregivers' acts.

The SLI children do not have the same verbal, linguistic abilities as the

NLA children in order to elaborate on the caregivers' verbal acts. The SLI children seem to compensate for this deficiency by using more nonverbal acts and backchannels. The high number of nonverbal acts and backchannels in the SLI dyads can also be related to the free-play situation studied. There is also a general significant increase in the average number of caregivers' backchannels ('Session Effect') in the SLI dyads. The SLI children gradually use more nonverbal acts from session 1 to 9. The increasing average number of the caregivers' backchannels and the decreasing average number of their verbal acts indicate that the caregivers of the SLI children create fewer opportunities for the child to respond coherently. The caregivers in the SLI dyads seem to change their language performance by reducing the number of verbal acts according to the predominant use of the nonverbal code by the SLI children. The language behaviour of the caregivers in the SLI dyads gradually tends to become more nonverbal and oriented to the children's activities. The average number of verbal acts of the caregivers and of the children exchanged in NLA dyads is much higher than in the SLI dyads (section 7.1). In contrast to the SLI dyads, the caregivers and the children in the NLA dyads try to cohere to each other's behaviours in a more verbal way.

The linguistic abilities of the NLA children indicate that they do not have to rely primarily on the use of (verbal) backchannels in order to cohere to preceding acts of themselves or of the conversational partner.

In the NLA dyads the caregivers and the children are similar in the way they use verbal acts to create discourse coherence. This can be attributed to a more equal conversational status between the NLA children and their caregivers than between the SLI children and their caregivers. The caregivers in the NLA dyads therefore do not need to adjust their verbal language performance because of another language code and level of performance used by the children.

7.2.2. Ellipsis

Hypothesis:

SLI children use more ellipsis than NLA children (rejected)

It is often hypothesised that linguistic immaturity (as in SLI) leads to an increased use of ellipsis, since ellipsis assumes that the necessary structure for a current utterance is recoverable from a previously used stretch of language (Fine, 1978; Van Ierland, 1982; Lasky & Klopp, 1984). On the other hand ellipsis is seen as an adequate and efficient means to participate in conversations and to create discourse coherence (Halliday & Hasan, 1976). Thus, a large amount of ellipsis in the language of SLI children can be seen as an indication of grammatical immaturity but also as an indication of the willingness to participate actively in discourse (Van Ierland, 1982). The data found in this study do not confirm the hypothesis that the SLI children use a higher average amount of ellipsis than the NLA children (Table 6.2.). The data indicate a 'Session Effect' for ellipsis used by the SLI children. This 'Session Effect' is mainly caused by the various dips and peaks in the average amount of ellipsis from session 1 to 9 (because of extreme indivual effects at the sessions 4 and 7).

There is a (non significant) decrease in the amount of ellipsis used in the subsequent sessions by the NLA children. Both groups show a gradual (non significant) decrease in the average amount of ellipsis used by the NLA children and an irregular pattern in the average amount of ellipsis used by the SLI childrren. The discourse involvement of the SLI children gradually becomes more nonverbal, obviously reducing the need to use ellipsis for creating coherent discourse. This explains that the difference in the amount of ellipsis used by the SLI and the NLA children is small and not significant.

7.2.3. Mean Length of Turn (MLT)

Hypothesis:
Caregivers of SLI children use longer turns than caregivers of NLA children
(rejected)

The data on 'turn length' in this study do not give sufficient support for the hypothesis stating that the caregivers of the SLI children use longer turns than the caregivers of the NLA children (Table 6.2.). The main reason for this is that the 'Mean Length of Turn (MLT)' is based on verbal as well as on nonverbal acts of each partner. In other studies concerning turn-exchanges in adult-child discourse and MLT measures, the turns and the MLT are primarily defined from the speaker's point of view. Nonverbal, turn-changing behaviour is rarely considered (Conti-Ramsden & Friel-Patti, 1984; McTear, 1985). The data and figures presented in the sections 6.2.1. and 6.2.3. indicate that the NLA and the SLI children and their caregivers use relatively short turns in their conversations. A small MLT mostly indicates an active, superficial exchange of information or the use of acts to manipulate the discourse. The kind of play-activities in which the caregivers and the children of the current study are involved are characterised by:
- nonverbal activities
- short, regulative turns (in order to manipulate the setting)
- backchannels (in order to comment on current activities)
- short conversational themes (due to the rapidly changing activities in play-situations)

(see also Lewis & Rosenblum, 1977; Schachter, 1979; Wells, 1985).

These characteristics partly explain the similarities in MLT for the SLI and the NLA dyads. Another reason for the absence of significant differences in

the MLT for the NLA and the SLI dyads can be found in the segmentation of nonverbal acts as defined in the present study.
The definition of a 'nonverbal act' influences the definition of a 'turn'.

What sort of nonverbal acts constitute a conversational turn? (For a definition of 'nonverbal act', see section 4.2.2. and for a definition of 'turn', see section 4.2.4.). Most studies on turn-taking in adult-child discourse do not address the issue of nonverbal behaviours. Therefore it is difficult to compare the results of nonverbal acts and MLT in the study at hand with other studies on turn-taking. Future research should explore in more depth the relationship between verbal and nonverbal acts and turn-taking.

7.2.4. Initiations Adult
Hypothesis:
Caretakers of SLI children initiate themes more often than caretakers of NLA children (rejected)
The hypothesis stating that the caregivers of the SLI children initiate discourse themes more often than the caregivers of the NLA children is not confirmed by the data of this study (Table 6.2.).

This study revealed significant 'Session Effect' for the caregivers' initiations, elaborations and reinitations in the SLI dyads.

In the NLA dyads the caregivers generally initiate conversational themes more often (not significantly) than the children. In the SLI dyads the children generally initiate conversational themes more often (Tables F2 and F4). In Wells' (1985;113) longitudinal study it appeared that the NLA children initiated about two-thirds of the sequences in conversation. The present study did not obtain that result.

In contrast to Wells' study, the current study used video-recorded language samples with equal consideration of verbal and nonverbal acts of both adult and child. Wells used verbal (audio-recorded) language data of the child mainly and adult utterances were only considered when they immediately preceded a child's utterance. Nonverbal initiations of the caregiver and adult-initiations which do not immediately go before a child's utterance were not counted in Wells' study. The study at issue did count all verbal and all nonverbal initiations of the caregiver.
This seems to explain the contrastive results in the study at issue (see the Figures 6.5a,b):
- in the NLA dyads the caregivers initiated the conversations more often than their children
- in the SLI dyads the children initiated conversations more often than their

caregivers, except for the first two sessions. These child-initiations are mainly nonverbal (see section 7.2.8.).

This indicates that the caregivers in the SLI dyads are not as active as their children and the caregivers in the NLA dyads in initiating conversational themes and in determining the discourse organisation.

7.2.5. Reinitiations Adult

Hypothesis:
Caretakers of SLI children reinitiate themes more often than caretakers of NLA children (confirmed)

Within IACV three kinds of initiations are scored: introductions of new themes, elaborations, and reinitiations of earlier introduced themes (section 4.3.1.).

Reininitiations used by the caregivers occurred significantly more often in the SLI dyads (Table 6.2.). Reinitiations generally function as a means to ensure the attention of the partner, to continue an earlier interrupted conversational theme, to enforce a specific discourse theme, or to elicit a specific response.

The caregivers in the SLI dyads try to get the explicit attention of the SLI children in order to evoke a specific reaction or to elaborate on an earlier introduced theme. This they do more often than the caregivers of the NLA children. This also explains, at least partly, the (not significantly) higher incidence of elaborations used by the caregivers of the SLI children, compared with the caregivers of the NLA children.

Generally the subsequences introduced by the SLI children are longer than the subsequences introduced by the NLA children (section 6.4.2.), mainly because of a general (non significant) tendency of using elaborations by the caregivers in the SLI dyads (as discussed in section 7.4.2.). The general trend of the caregivers in the SLI dyads to use more elaborations indicate that they are trying to create a framework for contingent interaction in order to compensate for the SLI child's verbal deficiencies, which limit the child's ability to elaborate on a theme (see also Bloom et al., 1976; Lasky & Klopp, 1982; McTear, 1985).

Extreme, individual scores of the caregivers at session 4 and 7 cause a 'Session Effect' in the SLI dyads. Besides these extreme, individual scores there is a general decrease in the caregivers' elaborations in the SLI dyads and a general (non significant) increase in the NLA dyads. Both groups of caregivers (in the NLA and in the SLI dyads) eventually arrive at similar frequency levels for elaborations at session 9. This indicates that the number of elaborations by the caregivers in the SLI dyads gradually reaches the 'normal' level needed for (normal) contingent discourse. The decrease in elaborations of the caregivers in the SLI dyads can be related to the gradual improvement in verbal skills of the SLI children from session 1 to session 9 (section 5.2.2.). It is interesting to relate these results to the findings reported in Lasky and Klopp (1982) where the use of expansions (elaborations) is significantly related to the NLA children's stage of language development and chronological age. For the SLI children a significant correlation is found between the number of the caregivers' backchannels (see section 7.2.1.) and the self-repetitions (see section 7.2.6.).

7.2.6. Self-related Imitations Adult
Hypothesis:
Caregivers of SLI children use more self-repetitions (or self-related imitations) than caregivers of NLA children (confirmed)

The hypothesis stating that the caregivers of the SLI children use more self-repetitions than the caregivers of the NLA children is confirmed by the data of this study. The study reveals a significant 'Between Groups Effect' for the caregivers' self-repetitions (Table 6.2.). The caregivers in the SLI dyads continue to use more self-repetitions than the caregivers in the NLA dyads from session 1 to 9 (Table F2).

There is a general and significant decrease in the caregivers' self-repetitions in the SLI dyads. The average number of the caregivers' self-repetitions in the SLI and NLA dyads gradually reaches a similar frequency level (about 2% of the total acts used) at session 9 (see Table F2 and the Figures 6.6a,b). Caregivers' self-repetitions generally provide a framework for contingent discourse and verbal learning (Lasky & Klopp, 1982; Roth & Spekman, 1984; McTear, 1985) Thus, it is to be expected that the average number of the caregivers' self-repetitions decreases as the child's verbal skills improve (see section 5.2.2. for the improvement of verbal skills). By inspection of the language production measures (partly indicated in chapter 5) and the figures for self-repetition, this expectation seems to be confirmed. A thorough description of the morpho-syntactic data in the IACV study should reveal more evidence for this reasoning.

7.2.7. Partner-related Imitations Child
Hypothesis:
SLI children use more imitations of preceding adult acts than NLA children
(rejected)

Children seem to learn to engage in discourse by first using exact imitations of the immediately preceding adult utterances and, later, by gradually expanding the caregivers' preceding utterances (Bloom et al, 1978; Bloom & Lahey, 1978).

The hypothesis concerning partner-related imitations is not confirmed by the data of this study (Table 6.2.). The SLI children tend to use more imitations of preceding adult utterances than the NLA children, although this difference is not significant (Figures 6.7a,b).

The SLI children seem to be less concerned with the acts of their caregivers. This can lead to incoherent, self (child)-directed discourse situations. The relatively low incidence of the caregivers' verbal acts in the SLI dyads (as discussed in section 7.1) reduces the opportunities of the SLI children to use partner-related verbal imitations.

7.2.8. Nonverbal Initiations

Hypothesis:
Initiative behaviour of SLI children is mostly nonverbal (confirmed)
The hypothesis that the SLI children's initiations are mostly nonverbal, is confirmed by the data of this study (Table 6.2.). The study also revealed a significant 'Session Effect' for the SLI children's nonverbal initiations. As indicated earlier, the SLI children gradually become more active in the manipulation of the discourse situation and the discourse organisation. The major part of the language performance of the SLI children is nonverbal. The increasing and continuous high number of nonverbal acts used by the SLI children explains the high and increasing number of nonverbal initiations of the SLI children (Friel-Patti & Conti-Ramsden, 1984; McTear, 1985). Figure 6.8b. indicates the growing number of nonverbal initiations of the SLI children, indicating their increasing influence and involvement in the discourse organisation.

7.2.9. Summary

The SLI children's verbal and linguistic deficiencies seem to limit their abilities to elaborate verbally on current conversational themes. Thus, they significantly use more nonverbal acts, nonverbal initiations and backchannels than the NLA children in order to compensate for their lack of verbal and linguistic abilities.

The caregivers in the SLI dyads try to create a coherent discourse more explicitly and actively than the caregivers in the NLA dyads. For this purpose the caregivers of the SLI children significantly use more reinitations of conversational themes and self-repetitions. Reinitations and self-repetitions are generally used to provide a framework for contingent discourse and verbal learning. These strategies help the SLI children to become more verbally and actively involved in discourse. The caregivers in the SLI dyads also use other strategies to establish coherent discourse more frequently than the caregivers in the NLA dyads, but the differences between both groups are not significant (e.g. initiations and elaborations). The language behaviour of the caregivers in the SLI dyads gradually becomes more nonverbal and oriented to the children's activities. The SLI and the NLA children do not differ significantly in the number of ellipses and partner-related imitations. The frequencies of these categories are low for both groups. Both groups of caregivers and children do not differ in the Mean Length of Turn. This result is mainly due to the kind of communicative interaction studied and the segmentation rules used.

7.3. Child Adjusted Register (CAR)

7.3.1. Mean Length of Utterance (MLU)
Hypothesis:
Compared with NLA children SLI children receive a language input characterised by a smaller MLU (rejected)
The data of this study do not confirm the hypothesis that the SLI children, compared with the NLA children, receive a language input characterised by a smaller average MLU (Table 6.3.).

The average MLU of the caregivers' utterances in the SLI dyads tends to increase (not significantly) from about 4.49 at session 1 to 5.33 at session 9 (Table F3). The average MLU of the caregivers' utterances in the NLA dyads takes varying values between 4.99 and 5.65 from session 1 to 9 (Table F3). The difference in the average MLU of the caregivers and the children in the SLI dyads from session 1 to 9 is higher than the difference in the average MLU of the caregivers and the children in the NLA dyads (see also Table F3 and Figure 6.9). This implies that the caregivers of the SLI children, compared with the caregivers of the NLA children, do not seem to adjust their language input (in terms of MLU) more explicitly to the level of language performance of the SLI children (see also section 5.2.2.) (Fey et al., 1981; Leonard, 1986). The average MLU of the caregivers' utterances in free-play sessions seems to have an upper limit of about 4.5. Below that average MLU utterances probably become difficult to understand and cannot function as clear frames for language and communication learning.

The non-significant differences in the MLU of the caregivers of the SLI and the NLA children is in contrast with the findings reported in Lasky and Klopp (1982). The nearly similar MLU of the caregivers in the SLI and NLA dyads is probably related to the specific kind of acts used in free-play sessions (e.g. elicitations, commands, requests) (Lewis & Rosenblum, 1977).

7.3.2. Self-related Imitations Adult
Hypothesis:
Compared with NLA children SLI children receive a language input consisting of more self-repetitions by the caregiver (confirmed)

The hypothesis stating that the caregivers of the SLI children use more self-repetitions than the caregivers of the NLA children has already been discussed in section 7.2.6.. The caregivers in the SLI dyads use significantly more self-repetitions than the caregivers in the NLA dyads. Self-repetitions of caregivers function as facilitation for language learning. The caregivers' self-repetitions can serve as conversational repair strategies (rephrasings) to prevent misunderstandings, justifications, 'attention preservers' and frameworks for learning to use contingent utterances and to ease the identification of turn-transition points (Bloom et al., 1976; Wells, 1981; Lasky & Klopp, 1982, McTear, 1985). The caregivers of the SLI children use more self-repetitions (especially in the early sessions) than the caregivers of the NLA-children in order to create a facilitative framework for the children to learn to create coherent discourse and to compensate for the SLI child's verbal and linguistic deficiencies to elaborate on conversation themes.

7.3.3. Requests for Information Adult
Hypothesis:
Compared with NLA children SLI children receive a language input consisting of more requests for information from the caregiver (rejected)

The hypothesis stating that the SLI children receive a language input consisting of more requests for information (except clarification requests, see section 7.6.3) than the NLA children is not confirmed by the data of this study (Table 6.3.). There is a significant 'Session Effect' for the caregivers' requests for information in the SLI dyads, indicating that the caregivers' average number of requests for information gradually decreases from session 1 to session 9 (see the Tables 6.3. and F3). Generally the caregivers in the SLI dyads tend to use more requests for information than the caregivers in the NLA dyads. Requests for information are mostly used to elicit extra information from the conversational partner about ongoing activities or topics previously discussed. It is often pointed out that SLI children have difficulty in grasping essential information from the context and the non-linguistic situation (Fey & Leonard, 1978; Schwabe et al., 1986; Conti-Ramsden, 1988). Caregivers try to assist their SLI child in gathering the necessary information from the context and the situation by requesting additional information. This presupposes a direct mutual discourse involvement of the caregiver and the child. According to the data concerning turn-taking discussed above, the SLI child seems to be more involved in his own activities. Thus, the caregiver needs to react to the child's activities in order to draw him into a conversational exchange of information and into sharing his activities with the caregiver. This leads to a discourse situation which is primarily child-oriented and organised from the caregiver's perspective (as discussed in several studies on topic manipulation in conversations with NLA children; e.g. Keenan & Klein, 1975; Brinton & Fujiki, 1984; Foster, 1986). Information requests seem to have a minor function in this kind of one-sided discourse orientation.

7.3.4. Summary
The caregivers in the SLI and in the NLA dyads do not differ significantly in the Mean Length of Utterance (MLU), although the difference in MLU between the SLI and the NLA children is significant. The average MLU of the caregivers' utterances in the free-play sessions studied are probably bound to a demarcation level of about 5. Below this level, utterances of the caregivers probably become difficult to understand for the children. The caregivers' adjustments in the SLI dyads are more clearly presented in the number of self-repetitions and information requests used per session. In the SLI dyads caregivers use significantly more self-repetitions than the caregivers in the NLA dyads in order to establish a coherent discourse situation. The caregivers in the SLI dyads differ from the caregivers in the NLA dyads in the number of information requests used per session. The number of caregivers' information requests decreases from session 1 to 9 in the SLI dyads, probably because of the number of nonverbal and self-directed acts of the SLI children. These acts do not necessitate and elicit a high number of information requests from the caregivers.

7.4.   Topic and Theme Management

7.4.1. Introduction New Theme
   Hypothesis:
   Compared with NLA Dyads, SLI Dyads are characterised by more 'introductions of new themes' used by the children (rejected)
The hypothesis stating that in the SLI dyads more child-initiated 'introductions of new theme' are used than the NLA dyads is not confirmed by the data of this study (Table 6.4.). This finding is in contrast with the suggestion that caregivers of SLI children have to initiate dialogue more often than caregivers of NLA children in order to create coherent discourse (Conti-Ramsden & Friel-Patti, 1988).

However, the non-significant finding in the current study gets evidence from the study of Keenan and Schieffelin (1976). In Keenan and Schieffelin's study it was found that children at early language acquisition stages (according to Brown, 1973) do not introduce conversational themes frequently and adequately for reasons of limited attention span, distractability and failures to grasp the essential information from the preceding utterances (context) and the non-linguistic situation. A similar situation is seen in the current study, in which the SLI children are in language acquisition stages comparable with the NLA children in Keenan and Schieffelin's study.

The average number of introductions of new themes is relatively low compared with the average number of elaborations (Table F2). Introductions of new conversational themes seldom occurred in the free-play sessions of the current study, mostly because the conversations were goal-directed and primarily oriented towards the play-materials available at the play-ground.

There is a significant 'Session Effect' for the mean number of themes introduced by the caregivers and the children in the SLI dyads. This 'Session Effect' is to be mainly attributed to extreme individual effects at the sessions 2, 5 and 8.

7.4.2. *Mean Length of Subsequence (MLS)*
Hypothesis:
The Mean Length of Subsequence (or Theme) of caregivers in SLI dyads is smaller than the Mean Length of Subsequence of caregivers in NLA dyads
(rejected)

The Mean Length of Subsequence (MLS) is related by definition to the average number of initiations of each partner per session (section 4.2.5.). The hypothesis stating that caregiver-initiated subsequences in the SLI dyads are shorter than caregiver-initiated subsequences in the NLA dyads is not confirmed by the data of this study (Table 6.4.). The reported significant 'Session Effect' concerning the MLS of the caregivers and the children in the SLI dyads is mainly caused by the steep increase in MLS from session 7 to 9. This indicates that the subsequences gradually become longer as the child's verbal skills grow (section 5.2.2.). The frequency pattern of the average MLS of the caregivers in the SLI dyads closely follows the frequency pattern of the average MLS of the SLI children (see the Figures 6.11a,b). It has already been discussed (sections 7.2.4. and 7.3.2.) that the high average number of the caregivers' self-repetitions in the SLI dyads helps the SLI child to become more verbally and actively involved in creating discourse coherence. The data on Mean Length of Subsequence (MLS) do support this suggestion (see also Barnes et al., 1983). Generally, child initiated subsequences in the SLI and in the NLA dyads tend to be longer (having a higher average Mean Length of Subsequence (MLS)) than the subsequences initiated by the caregivers (see the Figures 6.11a,b). The data on the average Mean Length of Subsequence (MLS) of the caregivers also indicate that they seem to help the children add new information to the conversational theme by using various strategies (e.g. elaborations, backchannels and self-repetitions). These strategies lengthen the subsequences (which occur more frequently in the SLI dyads) (McTear, 1985).

7.4.3. Partner-related Imitations Child
Hypothesis:
SLI children use more partner-related imitations than NLA children (rejected)
Children seem to learn to engage in discourse by first using exact imitations of the immediately preceding adult utterances and after that by gradually expanding the caregivers' preceding utterances (Bloom et al, 1978; Bloom & Lahey, 1978). The hypothesis concerning partner-related imitations is not confirmed by the data of this study (see also section 7.2.7., Table 6.2.). The SLI children tend to use more imitations of preceding adult utterances than the NLA children, although this difference is not significant (Figures 6.7a,b). The SLI children seem to be less concerned with imitating the verbal acts of their caregivers, probably because of their verbal deficiency and the nonexistent explicit need to verbally imitate the caregivers' utterances in the free-play sessions studied. Partner-related imitations function as facilitators in the learning of turn-taking and the functional use of contingent acts for the language acquiring child, especially during the early language stages 1 and 2 (see section 3.2.1.) (Brown, 1973; Bloom et al., 1976; Barnes et al.,1983). In light of the hypotheses discussed earlier, the absence of a significant 'Between Groups Effect' in the use of partner-related imitations can be seen as an indication of the SLI child's primary concern with his own activities and less with active partner-related activities. In both the NLA and the SLI dyads, the caregivers continue to use more partner-related imitations from session 1 to 9 than their children. However, the caregivers in both groups do not differ significantly in the use of partner-related imitations. These imitations mostly function as confirmations, prompts or requests for expansions from the child. This indicates that the caregivers in the SLI dyads react to their children's acts in order to gain (or regain) their attention and active discourse involvement, just as the caregivers of the NLA children.

7.4.4. Self-related Imitations Adult
Hypothesis:
Caregivers of SLI children use more self-related imitations than caregivers of NLA children (confirmed)
As already discussed in the sections 7.2.6. and 7.3.2. the hypothesis stating that the caregivers of the SLI children use more self-related imitations than the caregivers of the NLA children is confirmed by the data of this study (Table

6.4.). Self-related imitations function as devices for expanding a conversational theme (such as rephrasings, restatements, prompts, expansions). Through the use of self-related imitations, backchannels, elaborations and reinitiations the caregivers in the SLI dyads are trying to gain the children's conversational attention and to stimulate their willingness to become more verbally and actively involved in reciprocal discourse activities.

7.4.5. Summary

The majority of the hypotheses concerning 'Topic and Theme Management' is not confirmed. The main reason for discarding most of the hypotheses is that they all originated from studies in caregiver-child interaction in which only one partner (the caregiver or the child) or only verbal turn-exchanges were investigated. The relation between verbal and nonverbal acts and the reciprocity of the acts used by both partners are the most important aspects in discussing the results of the study at issue. These aspects also form the principal distinction with other related studies on caregiver-child discourse. Consequently this leads to differing results and alternative conclusions. As already indicated in the earlier sections of this chapter, the caregivers in the SLI dyads try more explicitly to verbally expand on the children's activities (e.g. by using self-repetitions, elaborations). In the SLI dyads subsequences become significantly longer during the last sessions, which seems to be related to the gradual improvement of the children's verbal skills.

7.5. Communication Functions

7.5.1. Requests for Information Adult and Child

Hypothesis:

SLI children use fewer requests for information than NLA children (rejected)

The hypothesis stating that the SLI children use fewer requests for information than the NLA children is not confirmed by the data of this study (Table 6.5.). In Wells' report of the Bristol longitudinal language development project, the average number of requests in Representationals (for definition see section 4.3.7. and Appendix C, section 5) remains rather low in frequency (Wells, 1985, Table A3). The data found in the present study show similar frequency patterns, slightly higher for the NLA children than for the SLI children. The absence of a significant difference between the SLI and the NLA children seems to be related to the fact that nonverbal requests were also incorporated here. The SLI children seem to compensate their verbal deficiency for requesting information by using nonverbal requests too (McTear, 1985).

7.5.2. Communication Functions Adult and Child
Hypothesis:
Caregivers of SLI children use more Controls and Tutorials than caregivers of NLA children (rejected)

The hypothesis stating that the caregivers of the SLI children use more Controls and Tutorials than the caregivers of the NLA children is not confirmed by the data of this study (Table 6.5.). Although there are no significant differences in the numbers of Controls and Tutorials of the caregivers in both groups, the study revealed interesting (though non-significant) differences in the various sub-categories of Controls and Tutorials used by the caregivers (Appendix E, Tables E-1, E-2). The caregivers in the SLI dyads tend to show a higher mean number of Requests (as was also indicated earlier in section 7.3.3.), Wantings/Intends, Prohibitions and Refuses/Rejects in the Control Function than the caregivers in the NLA dyads. The caregivers in the SLI dyads also tend to show a higher average number of all sub-categories in the Tutorial Function (for definitions of the Communication Functions and their sub-categories, see Appendix C). These differences were not significant when compared with the NLA group, mainly attributed to considerable individual fluctuations in the standard deviations of the SLI group (Table F5). The general trend of these results is that the caregivers in the SLI dyads seem to try more intensively to create a discourse situation in which the SLI children become more active and verbal. Although the caregivers in both groups do not differ significantly in the average number of Controls and Tutorials they tend to differ in the kind of acts within Controls and Tutorials.

The significant 'Session Effect' for the Procedurals used by the caregivers in the SLI dyads is primarily caused by the increase in Calls and Contingent Queries from session 1 to 5. The significant 'Session Effect' for the Representationals in the group of SLI children and the nearly significant 'Session Effect' in the group of NLA children is mainly the result of a general increase in the use of Statements from session 1 to 9.

The significant 'Session Effect' for the Expressives used by the SLI children is mainly the result of an overall increase in the average number of Expressives (mainly Exclamations and Accompaniments to Action) from session 1 to 7. Expressives are mostly used in combination with nonverbal acts. This supports the suggestion that the SLI children are more involved in their own activities than in the activities of their caregivers.

All these results tend to support the above stated supposition that the caregivers in the SLI dyads actively try to get the children's attention in order to establish an interaction situation in which coherent discourse becomes possible.

7.5.3. Summary
The current study did not reveal significant differences in the number of Communication Functions used by the caregivers and the children in either groups. The most interesting finding is the considerable difference in the variety of communication acts used within each category of Communication Function. In contrast to findings in other studies concerning adult-child discourse it is found that the SLI and the NLA children do not differ significantly in the use of 'Information Requests', 'Controls', 'Representationals' etc., mainly because of the inclusion of verbal and nonverbal acts in the study at issue.

There are some interesting 'Session Effect' in the SLI dyads for the caregivers' use of 'Information Requests' and 'Procedurals' and for the children's use of 'Controls', 'Expressives', 'Representationals' and 'Procedurals'. In the NLA dyads only one (nearly) significant 'Session Effect' occurred for the children's use of 'Representationals'. All these results support the supposition that the caregivers in the SLI dyads try more explicitly to get the children's attention in order to establish an interaction situation in which coherent discourse becomes possible. The communication acts used by the SLI children in the several categories of Communication Functions indicate their preoccupied involvement in self-directed activities.

7.6. Communication Breakdowns and Repairs

7.6.1 Unfinished Verbal Acts
   Hypothesis:
   SLI children use more unfinished verbal acts than NLA children (rejected)
The hypothesis stating that the SLI children use more unfinished verbal acts than the NLA children is not confirmed by the data of this study (Table 6.6.). Although there is no significant difference in the number of unfinished verbal acts used by the NLA and the SLI children, the SLI children generally tend to use more unfinished utterances. The frequencies of unfinished verbal acts used in the SLI and NLA dyads are too small to discuss these observed (minimal) differences. The low frequency of unfinished verbal acts is related to the small number of verbal acts used by the SLI children.

7.6.2. Parallel Talk
   Hypothesis:
   Compared with NLA dyads, SLI dyads are characterised by a higher average number of parallel talk (rejected)
The hypothesis stating that the SLI dyads are characterised by a higher inci-

dence of parallel talk than the NLA dyads is not confirmed by the data of this study (Table 6.6.). The expected higher amount of parallel talk in the SLI dyads was based on the supposition that the caregivers and the children are mainly involved in their own activities and do not cooperate successfully in discourse. This assumption was mainly based on the observed verbal and linguistic deficiencies of the SLI children, which seems to limit their ability to initiate cooperative discourse. This obviously can result in a limited number of occasions to which the caregiver can react. The data of the current study did not support this supposition.

As mentioned in the previous sections, the data of this study indicate that the SLI child mostly uses nonverbal acts and primarily is concerned with his own activities. Thus, the occasions which can lead to parallel talk are limited.

7.6.3. Adult Clarification Requests
Hypothesis:
Caregivers of SLI children use more clarification requests than caregivers of NLA children (confirmed)

The hypothesis stating that the caregivers of the SLI children use more clarification requests than the caregivers of the NLA children is confirmed by the data of this study (Table 6.6.). The study also revealed a significant 'Session Effect' for the caregivers in the SLI dyads. The observed 'Session Effect' is mainly caused by a decrease in clarification requests from session 7 to 8. The 'Between Groups Effect' is mainly caused by the higher average number of clarification requests by the caregivers in the SLI dyads.

In light of the earlier discussed results the caregivers of the SLI children are trying more often to become involved in the discourse activities initiated by their children. Clarification requests are an efficient means to elicit the partner's attention and reactions (McTear, 1985; Brinton et al.,1986; Prather et al., 1989).

7.6.4. Adult Corrections of Previous Act
Hypothesis:
Caregivers of SLI children use more corrections than caregivers of NLA children (confirmed)

Similar to the use of clarification requests, the caregivers of the SLI children use significantly more corrections than the caregivers of the NLA children. In general, the use of corrections by the caregivers of the SLI children decreases significantly from session 1 to 9, indicating that corrections gradually are seen as a less useful instrument to influence and regulate the child's activities.

In fact corrections also function as communication breakdowns, because they disrupt a smooth information exchange (McTear, 1985). Most of the caregivers' corrections are verbal. As indicated in section 7.1., the average number of the acts and more specifically of the verbal acts of the caregivers decreases from session 1 to 9.

This decrease in verbal acts also partly explains the decrease in corrections. Another explanation for the significant decrease in corrections probably lies in the gradual improvement of the SLI children's verbal abilities (section 5.2.2.).

7.6.5. Consecutive Initiations
Hypothesis:
SLI dyads show a higher incidence of consecutive initiations than NLA dyads (rejected)

Consecutive initiations are, similar to parallel talks an indication of communication breakdowns. They show possible mismatches in conversational intentions of the partners. The hypothesis stating that the SLI dyads are characterised by more consecutive initiations than the NLA dyads is not confirmed by the data of this study (Table 6.6.). A close examination of Table F6 shows that consecutive initiations tend to occur more often in SLI dyads than in NLA dyads. The differences between the SLI and NLA dyads are not significant mainly because of the large standard deviations. In fact the SLI dyads generally indicate a higher average number of consecutive initiations, this seems to imply that there is a general tendency in the SLI dyads signifying that the caregivers and the children appear to introduce conversational themes subsequently. Both the caregivers and the children tend to regulate each other's behaviour in order to manipulate the discourse situation in favour of their own intentions.

7.6.6. Faulty Responses
Hypothesis a.:
Caregivers in SLI dyads show more faulty responses than the caregivers in NLA dyads (confirmed)

Hypothesis b.:
SLI children show more faulty responses than the NLA children (rejected)

The hypothesis stating that the caregivers and the children in the SLI dyads show more faulty partner-related responses than caregivers and the children in the NLA dyads is confirmed by the data of this study, as far as the data of the caregivers are concerned (Table 6.6.). The study reveals that there is a significant 'Between Groups Effect' for the caregivers' faulty responses. Faulty responses refer to pragmatically incorrect responses which do not elaborate on a previous question, request or partner-related nonverbal activity (see section 4.4.3.). The caregivers in the SLI dyads tend to use refusals*, rejection* and prohibitions* of children's intentions more often than the caregivers of the NLA children (see also Appendix E, Tables E-1 to E-5). Not all of these negations are pragmatically incorrect, but the major part of these acts leads to communication breakdowns and consequently are coded as pragmatically incorrect. These incorrect responses occur less frequently in the NLA dyads.

The results of incorrect responses of the caregivers in the SLI dyads support the assumption that the caregivers of the SLI children influence and redirect the children's activities, discussed previously. Bedrosian et al. (1988) came to similar conclusions in their study of turn-taking violations in adult-child conversations. Contrary to their expectations the authors observed that the caregivers exhibited more turn-violations than their children. The role of the caregivers, therefore, appeared to be closely aligned with Control. The observed significant 'Session Effect' for incorrect responses of the NLA children is mainly the result of individual effects at session 7 and session 8.

The expectation that the SLI children will use more faulty responses than the NLA children is not confirmed by the data of this study. This result can partly be attributed to the SLI children's primary use of nonverbal acts related to self-directed activities (section 7.1. and 7.5.). Another aspect which can have influenced this result is the kind of interaction studied (free play sessions). Conversations in free play sessions are probably more directed to nonverbal activities and create less opportunities to respond.

7.6.7. Faulty Initiations

Hypothesis a.:
Caregivers in SLI dyads use more faulty initiations than caregivers in NLA dyads (confirmed)

Hypothesis b.:
SLI children use more faulty initiations than NLA children (rejected)

The hypothesis stating that the caregivers and the children in the SLI dyads show more faulty initiations than the caregivers and the children in the NLA dyads is confirmed by the data of this study, as far as the data of the caregivers are concerned (Table 6.6.). The caregivers in the SLI dyads use more faulty initiations than the caregivers in the NLA dyads. Faulty initiations are coded from the perspective of the receiver/hearer (e.g. explicitly made corrections, introductions of new themes and reinitations which disrupt the information flow, see section 4.4.3.). The caregivers of the SLI children probably use significantly more faulty initiations than the caregivers of the NLA children because they try to influence the children's activities in order to establish a more

*See for definitions of these sub-categories of Control Appendix C.

coherent discourse situation (as discussed earlier). The study also indicates significant 'Session Effect' for faulty initiations used by the caregivers in the SLI and the NLA dyads and used by the SLI children. The 'Session Effect' for the caregivers in the SLI and in the NLA dyads and for the children in the SLI dyads are mainly the result of individual effects at the sessions 3 and 6.

SLI and NLA children did not differ significantly in the use of faulty initiations.

By inspection of the data and figures concerning the kind of initiations and the relations of acts with previous acts used in the SLI and the NLA dyads, it seems that the SLI children's initiations tend to be more related to self-directed activities. The caregivers' initiations in the SLI dyads tend to be more related to the children's acts. The SLI caregivers' initiations which have the intention to direct the children's attention more explicitly to the conversational floor often disrupt the ongoing activities of the child. These initiations are coded as 'faulty initations'. This is probably the reason for finding a significant 'Between Groups Effect' for the caregivers' use of faulty initations and not for the SLI and the NLA children.

7.6.8. Summary
The study at issue revealed significant 'Between Groups Effect' for the number of clarification requests, corrections, faulty initiations and responses used by the caregivers. All these significant results are interpreted in the light of the earlier discussed strategies of the caregivers in the SLI dyads to regulate the discourse situation more explicitly. These strategies have the intention to help the SLI children to become more verbally and actively involved in discourse manipulation. All 'Session Effect' concerning the hypotheses of 'Communication Breakdowns and Repairs' are caused by individual effects in several sessions. There is no general increasing or decreasing trend from session to session.

The hypotheses concerning the use of 'Unfinished Verbal Acts', 'Parallel Talk' and 'Consecutive Initiations' were discarded, mainly explained by the inclusion of nonverbal acts and the relatively small number of verbal acts coded in this study.

7.7. Overall Summary of the Results

In the previous sections a discussion of the hypotheses stated in chapter 4 was given based on the results presented in chapter 6. The hypotheses all concerned discourse coherence or the way in which discourse coherence is estab-

lished or disturbed in caregiver-child conversations. This section gives an overall discussion of the links between the results considered in the preceding sections. The discussion focuses on the characteristics of the caregivers' and children's discourse behaviour in three main areas of the study at issue:
1. The Participation in Discourse Coherence
2. The Role of the Caregiver
3. Grammatical Development and Pragmatic Skills.

7.7.1. The Participation in Discourse Coherence
Verbal and Nonverbal Acts
The study at issue revealed significant differences in the number of nonverbal acts and initiations used by the SLI children and the NLA children. Besides this the SLI children significantly used more backchannels than the NLA children. The caregivers in the SLI dyads tend to use fewer verbal acts than the caregivers in the NLA dyads from session 1 to 9. The caregivers in the SLI dyads start using significantly more backchannels than the caregivers in the NLA dyads from session 1 to 9. The total number of acts performed by the caregivers in the SLI dyads tends to decrease from session 1 to 9. The SLI children initiate discourse more often than their caregivers. However, the SLI children's initiations are predominantly nonverbal and directed to their own activities. These results taken together indicate that the SLI children are active participants in discourse. They primarily use nonverbal acts and nonverbal initiations to manipulate discourse themes which are primarily oriented towards self-related play activities.

The SLI children's use of backchannels indicates that they are aware of being involved in a mutual discourse situation with their caregiver. Their linguistic and verbal deficiencies seem to limit their ability to become more actively involved in establishing a coherent conversation. Backchannels and nonverbal acts then serve as alternatives and compensation for these deficiencies. The caregivers of the SLI children are trying to become more involved in their children's activities. The caregivers gradually adjust their language performance to the predominantly nonverbal behaviour of the SLI children. The increased use of nonverbal acts, backchannels and reinitiations of the caregivers in the SLI dyads indicate their efforts to gradually take part in the children's activities. This strategy used by the caregivers in the SLI dyads seems to have the intention to gain the children's attention and to take their activities as a starting point for creating a coherent discourse situation. This explains also why the number of initiations of the SLI children tends to be higher than the number of the initiations of their caregivers. The general decrease in the number of acts used in the SLI dyads signifies that the role of the caregiver gradually becomes more closely aligned with that of control rather than facilitation. Superficially seen these results seem to be in contrast with the findings and suggestions made in related studies (Watson, 1977; Fey et al., 1981; Lasky & Klopp, 1982; Roth & Spekman, 1984; Scherer & Olswang, 1984; Dromi & Beni-Noked, 1984; Conti-Ramsden & Friel-Patti, 1984; Conti-Ramsden, 1988). It is expected from these studies that the caregivers of the SLI children will use more initiations and verbal acts than their children and the caregivers of the NLA children in order to create coherent discourse. In fact, the data in the present study support the active caregivers' role in creating a coherent discourse. However, the caregivers of the SLI children use another kind of strategy. At least, it takes two persons to create a coherent discourse. The caregivers in the SLI dyads therefore try to get more involved in the predominantly nonverbal activities of their children in order to get their attention and to change the self-directedness of the children's acts into a more partner-related orientation.

For that reason the caregivers try to elaborate more on the child-initiated activities instead of initiating new themes. The results related to the other hypotheses of the current study are in support of this interpretation. The main difference with the present study and other related studies in adult-child discourse is the consideration of reciprocal relations between the verbal and nonverbal acts used by both partners. The set-up of this study enables the observation of the mutual relationship between the acts of both partners in a longitudinal perspective.

Partner-related Imitations and Nonverbal Acts
The major part of the acts performed by the SLI children is nonverbal and is related to self-activities. Therefore, the acts used by the SLI children have weaker relationships with preceding adult acts than the acts used by the NLA children and occur less frequently. As suggested in several studies children seem to learn their roles in conversation by imitating (and thus creating relations with) the language behaviour of their caregivers (Bloom et al., 1976; Bloom & Lahey, 1978; Lasky & Klopp, 1982). The SLI children in this study are at Brown's language development stages 2 and 3 (see also section 3.2.1.). Thus, it is expected that the SLI children will use significantly more partner-related imitations than the NLA children in this study. The NLA children are at a much higher level of language development in which a high number of partner-related imitations is not expected anymore. The present study did not find significant differences in the number of partner-related imitations used by the SLI and the NLA children. The number of imitations used by SLI and the NLA children in the present study is small (see Appendix F, Table F2). This result was expected for the NLA children. As already indicated in the previous sections the acts used by the SLI children are mostly self-directed and non-

verbal. In accordance with these data it can be explained that SLI children will use fewer partner-related imitations than expected.

Some reasons for the counter-evidence found for the expected higher number of partner-related imitations in this study probably are :
- the kind of language used in free play sessions
- a combination of self-directedness and linguistic deficiency of the SLI children (which has to be worked out more precisely by using the morpho-syntactic data of the SLI children)
- the kind of utterances used in Control acts of the caregivers in the SLI dyads, intended to direct the children's conversational attention, probably are not ideal for getting imitated.

7.7.2. The Role of the Caregiver
This section will focus on the kind of specific characteristics of communicative interaction of the SLI children and their caregivers in order to identify the mutual relationships between features of the children's communicative abilities and specific features of the adult input. Some of these specific characteristics have already been discussed in the previous section.

Influences on the Mean Length of Subsequence
The current study did not find significant differences in the Mean Length of Subsequences of the NLA and the SLI dyads. Some studies on topic manipulation in conversations between the SLI children and their caregivers indicate that the child-initiated topics contain a larger number of utterances and turn-exchanges than the adult-initiated topics (Liles, 1985; Johnston, 1986; Prutting & Kirchner, 1987). Generally the child-initiated subsequences in the SLI and the NLA dyads of the study at issue tend to be longer than the caregiver-initiated subsequences. In fact the caregivers of the SLI and the NLA children do not differ in the number of acts used to elaborate on conversational themes initiated by the children. The caregivers of the SLI children talk in much the same way as the caregivers of the NLA children do (Lasky & Klopp, 1982).

The strategies used by the caregivers to elaborate on child-initiated subsequences help the children in becoming more involved in contingent discourse. However, the kind of acts used by the caregivers of the SLI and the NLA children to elaborate on child-initiated subsequences differs. The caregivers in the SLI dyads used self-repetitions significantly more often than the caregivers in the NLA dyads. This finding was also reported in Lasky and Klopp (1982). Self-repetitions were used more often by the caregivers in the SLI dyads in order to provide acknowledgement of a previous act or to supply some continuity or coherence in discourse (Keenan & Klein, 1975). Cross (1977) suggests that the majority of the caregivers' self-repetitions followed the child's failure to respond appropriately. This lack of response is probably responsible for the incidence of self-repetitions. As a consequence the number of self-repetitions should decrease with the child's improvement of verbal and linguistic abilities (Lasky & Klopp, 1982). The present study seems to support this expected decrease of the caregivers' self-repetitions when associated with the children's progress in verbal and linguistic abilities. The results reported by the studies considering the caregivers' use of self-repetitions support also the significant differences in the number of faulty and consecutive initiations between the caregivers of the SLI and the NLA children. Besides the use of self-repetitions, the caregivers in the SLI dyads also differ from the caregivers of the NLA children in their use of significantly more backchannels per session. Backchannels elaborate also on previously introduced themes and in fact stimulate the speaker or actor (the SLI child) to continue with talking or acting. Another (non significant) trend in which the caregivers of the SLI children differ from the caregivers of the NLA children concerns the higher proportion of nonverbal acts used. The caregivers in the SLI dyads try to become more engaged in the activities of their children in order to gain the children's attention and more active involvement in mutual discourse. For that reason the caregivers tend to use more nonverbal acts and fewer initiations in order to elaborate on child-initiated themes.

The effects of these caregiver strategies are discussed in several other studies (e.g. Bloom et al., 1976; Barnes et al., 1983; Scherer & Olswang, 1984; Lasky & Klopp, 1984). The findings of these studies support the stimulating effect of these strategies on language acquisition and discourse involvement discussed in the present study.

Faulty Initiations of the Caregivers
The caregivers in the SLI dyads try more explicitly to (re)gain the children's attention in order to cooperate and create coherent discourse than the caregivers of the NLA children. As a consequence this leads to initiations of the caregivers which disrupt the child's activities. Initiations which intrude ongoing activities were coded as 'faulty initiations'. This explains why the caregivers in the SLI dyads show a significantly higher number of faulty initiations and responses. It also explains the high number of consecutive initiations in the SLI dyads. Bedrosian et al. (1988) also reported that the mothers in their study exhibited more turn-violations than did their (NLA) children. The frequency of these turn-violations decreased significantly with the children increasing in age. The role of the caregivers, therefore, seems to be more closely aligned with that of control rather than facilitation. This aspect was also dis-

cussed in relation to the significantly high number of reinitiations used by the caregivers in the SLI dyads in the previous section. A similar trend in the decrease of consecutive initiations and faulty initiations and responses is seen in this study.

Communication Functions
According to Lasky and Klopp (1983) SLI children seem to have no difficulties in their performance of communication acts despite significant limitations in their morpho-syntactic abilities.

Most studies report a deficient use in the range of communication acts in addition to problems in formal linguistic skills (Fey & Leonard, 1983). Fey and Leonard's review of the literature indicates that SLI children reflect deficiencies in their ability to produce requests for action and information.

Keenan and Schieffelin (1976) reported a similar conclusion. The present study indicates that the SLI and the NLA children do not differ significantly in the number of communication functions used.

This seems to be in accordance with the data discussed by Lasky and Klopp (1983), although the present study did not present an in-depth description and discussion of the morpho-syntactic abilities of the SLI children (because of the main purpose of this study, which is more pragmatical oriented). By inspection of the data in section 5.2.2., the morpho-syntactic abilities of the SLI children seem to be limited compared to the NLA children.

The data relating to the number of communication acts used by the SLI children are in conflict with some of the findings reported in the Fey and Leonard study (1983). An examination of the reported studies concerning Communication Functions reveals that most of them do not consider nonverbal acts and thus do not code their communication function. The present study coded the Communication Functions of verbal as well as nonverbal acts. The major part of the counter-evidence in the findings of the current study probably is to be attributed to the coding of verbal and nonverbal acts. Although there are no significant differences in the numbers of Communication Functions between the children and between the caregivers of both groups, the study revealed interesting (though non-significant) trends in the varying use of sub-categories in Controls and Tutorials especially for the caregivers in both the SLI and the NLA dyads (the relevant data are presented in Appendix E, Tables E-1, E-2). The caregivers in the SLI dyads seem to use more Requests, Wantings/Intends, Prohibitions and Refuses/Rejects in the Control Function than the caregivers in the NLA dyads. The caregivers in the SLI dyads also tend to show more variety in the use of acts in all sub-categories of the Tutorial Functions. However, these differences were not significant when compared with the NLA group. These data tend to indicate that the caregivers in the SLI dyads are busy trying to create a discourse situation in which the SLI children can become more actively and verbally involved.

Although the caregivers in both groups do not differ significantly in the average number of total acts used in Controls and Tutorials, they tend to differ with respect to the use of different sub-categories within Controls and Tutorials. The significant 'Session Effects' for the Procedurals used by the caregivers in the SLI dyads, for the Representationals and Expressives in the group of SLI children seems to support the suggestion that the SLI children are more involved in their own activities than in the activities of their caregivers. All these data tend to support the above stated supposition that the caregivers in the SLI dyads actively try to get the children's attention in order to establish an interaction situation in which coherent discourse becomes possible.

Future research should consider differentiations of sub-categories of Communication Functions and should also code the communicative function of nonverbal acts. The relation between morpho-syntactic and communicative abilities of children deserves more attention in longitudinal studies of language development.

Clarification Requests
The present study revealed the use of significantly more clarification requests by the caregivers of the SLI children. The use of more caregiver clarification requests in the SLI dyads is also reported in other studies (e.g. Gallagher, 1977; Gallagher & Darnton, 1978; Brinton et al.,1986; Prather et al., 1989). These studies explain the use of significantly more clarification requests in SLI dyads as a strategy of the caregivers to repair communication breakdowns caused by the children's incorrect or inappropriate responses. A relation between the use of clarification requests and subsequent repairs is suggested in most of the studies (McTear, 1985). In the light of the interpretation of the results of the study at issue the use of clarification requests by the SLI caregivers is also motivated by their wish to regulate and redirect the activities of the children in order to create coherent discourse. Clarification requests seem to be an efficient means to elicit the child's attention and a partner-related response.

Corrections
The scarce number of studies concerning communication breakdowns and repairs in caregiver-child interaction and more specifically in conversations between SLI children and their caregivers suggest that the problems SLI children have with understanding and creating coherent discourse lead to an increase of misunderstandings (e.g. incorrect responses, clarification requests, corrections and self-repetitions) (e.g. Gallagher, 1977; Gallagher & Darnton, 1978; McTear,1985; Brinton et al.,1986; Prather et al., 1989). The use of signifi-

cantly more self-repetitions and clarification requests by the caregivers in the SLI dyads has already been discussed. The present study supports the findings of other related studies concerning communication breakdowns and repairs. The caregivers of the SLI children use significantly more corrections than the caregivers in the NLA dyads. Corrections are seen as a more explicit means to repair communication breakdowns than for instance clarification requests and self-repetitions.

The current study also indicated a significant decrease in the corrections used by the caregivers of the SLI children from session 1 to 9. This decrease seems to be related to a gradual improvement of the SLI children's verbal and linguistic skills (as indicated in section 5.2.2.). Due to the limited number of studies on this topic additional research is needed in order to further investigate the relationships between linguistic maturation of the children and a decrease in the number of communication breakdowns and consequently in corrections.

7.7.3. Grammatical Development and Pragmatic Skills

The SLI children were all selected primarily because of their difficulties in language production. At the beginning of the IACV project the SLI children had an average delay in language comprehension of about three months (compared with their chronological age) and an average delay in language production of about fourteen months (compared with their chronological age) (as indicated in section 5.2.2.). At the end of the IACV project the average language comprehension age of the SLI children seems to be at the appropriate level (compared with their chronological age). The language production of the SLI children improved considerably at session 9, but still seems to be delayed when compared with the language production of the NLA children. The results of the study reported in this book did not consider the possible morpho-syntactical difficulties the SLI children probably have with creating discourse coherence. However, several results discussed above referred to an improvement of verbal and linguistic skills of the SLI children (as can be derived from the Tables 5.2. to 5.4.).

Only a few studies considered the interesting relationship between grammatical and pragmatic development of NLA (Ochs & Schieffelin, 1983; McTear, 1985; Wells, 1985) and SLI children (Bloom & Lahey, 1978; Russo & Owens, 1982; Lasky & Klopp, 1982).

The following aspects were partly explained in the light of the SLI children's improvement of verbal and linguistic skills:
- the decrease of the caregivers' number of self-repetitions in the SLI dyads
- the decrease of the SLI children's number of nonverbal initiations from session 7 to 9

- the increase in MLU in the language production of the SLI children
- the increase in the Mean Length of Subsequence (MLS) in the SLI dyads
- the decrease of adult clarification requests in the SLI dyads
- the decrease of adult corrections in the SLI dyads.

Generally seen the SLI children in the present study do not seem to profit from their improved verbal and linguistic skills in becoming a more competent conversational partner.

A study more in-depth concerning the relation between grammatical development and these aspects of pragmatic development is planned and will be published. This publication hopefully will lead to a thorough discussion about this relationship.

7.7.4. *Implications for Further Research*
Interindividual differences of the SLI children
The frequency patterns of most language categories studied do not indicate a clear improvement in discourse involvement. There often are extreme fluctuations in the frequencies of the SLI children's data in session 1 and session 9, but the final result at session 9 does not always differ from the achievements at the first sessions. This observation leads to another general characteristic in the frequency data of the SLI children: the group frequencies of the SLI children (for each session) show extreme variations in Standard Deviations, indicating obvious interindividual differences between the SLI children. These extreme variations in Standard Deviations are not seen in the frequency data of the NLA children.

In a similar study of Feilberg (Norway) equivalent variations in Standard Deviations of the frequency data from SLI children have been found (personal note from Conti-Ramsden, 1990). Conti-Ramsden also mentioned (in a personal comment to the author) that these variations in Standard Deviations were found in her data. These observations indicate the need for more longitudinal case study designs in which SLI and NLA children are investigated in conversational settings in order to collect more data for explaining these extreme interindividual variations in SLI children.

Linguistic and conversational competence
The present study also revealed the necessity for more research in the way of creating discourse coherence in conversations with SLI children. This research should mainly focus on the relation between linguistic and conversational competence.

Comparative studies of language input to SLI and NLA children
This study also indicates the extreme importance of comparative research in the use of language input strategies (or Child Adjusted Register) with NLA and SLI children. The differences in the kind of communication acts used by the caregivers in the SLI and NLA dyads deserves more attention in future research. Besides interesting findings for further child language research the results of this study have important clinical implications for the communication training of SLI children and their caregivers. This issue will be described in a clinical report of the IACV project.

BIBLIOGRAPHY

Atkinson, M. (1979)
  Prerequisites for Reference.
  In: Ochs, E & Schieffelin, B.B. (eds.) *Developmental Pragmatics* New York: Academic Press, 229-249.
Austin, J.L. (1962)
  How to do Things with Words. Oxford: Oxford University Press, 1962.
Balkom, H. van; Van Blom, K.; Groeneweg-Bruckman, L. & Fasotti-Dumont, T. (1989)
  Interactie Analyse van Communicatieve Vaardigheden (IACV). Algemene handleiding voor transcriptie en het uitvoeren van IACV analyses. Hoensbroek: IRV/01 Doc. (89).
Balkom, H. van & Heim, M. (1990)
  The Methodological Challenge of Interaction Research in Augmentative and Alternative Communication (AAC).
  Issues Paper presented at the International Symposium on Research in AAC, Norra Latin City Conference Centre, Stockholm, Sweden, August 16 & 17, 1990.
Balkom, H. van & Welle Donker-Gimbrère M. (1988)
  Kiezen voor Communicatie. Een Handboek over de Communicatie van Mensen met een Motorische of Meervoudige Handicap. Nijkerk: INTRO.
Barnes, S.; Gutfreund, M.; Satterly, D. & Wells, G. (1983)
  Characteristics of adult speech which predict children's language development.
  Journal of Child Language, 10 (1983), 65-84.
Bates, E.; Camaionie, L. & Volterra, V. (1975)
  The acquisition of performatives prior to speech.
  Merrill-Palmer Quarterly, 21, 3, 205-226.
Bates, E. (1979)
  Language in Context: The Acquisition of Pragmatics.
  New York: Academic Press.
Bates, E. & MacWhinney, B. (1979)
  A functional approach to the acquisition of grammar.
  In: Ochs, E. & Schieffelin, B.B. (eds.), *Developmental Pragmatics.* New York: Academic Press.
Bateson, M.C. (1975)
  Mother-infant exchanges: the epigeneses of conversational interaction.
  In: Aaronson, D. & Rieber, R.B. (eds.) *Developmental Psycholinguistics and Communication Disorders, Annals of the New York Academy of Science* 263, 101-112.
Beaugrande, R. De (1980)
  Text, Discourse and Process. Towards a Multidisciplinary Science of Texts. Florida, Miami: First Longman edition.
Becker, J.A. (1988)
  The success of parents' indirect techniques for teaching their preschoolers pragmatic skills.
  First Language, 8, 23, 173-182.

Bedrosian, J.L.; Wanska, S.K.; Sykes, K.M.; Smith, A.J &
Dalton, B.M. (1988)
Conversational Turn-taking violations in mother-child interaction. *Journal of Speech and Hearing Research*, 31, 81-86.

Berninger, G. & Garvey, C. (1982)
Tag Constructions: Structure and function in child discourse.
Journal of Child Language, 9, 151-168.

Blank, M. & Franklin, E. (1980)
Dialogue with preschoolers: A cognitively-based system of assessment.
Applied Psycholinguistics, 1, 127-150.

Blau, A.F. (1986)
Communication in the back-channel: Social structural analysis of nonspeech conversations.
Unpublished doctoral thesis, University of New York, USA.

Bloom, L. (1970)
Language Development: Form and function in emerging grammars. Cambridge, Mass.: The MIT Press.

Bloom, L. & Lahey, M. (1978)
Language Development and Language Disorders.
New York: John Wiley & Sons.

Bloom, L.; Roscissano, L. & Hood, L. (1976)
Adult-child Discourse: Developmental interaction between information processing and linguistic knowledge.
Cognitive Psychology, 8, 521-552.

Blount, B.G. (1977)
Ethnography and caretaker-child interaction.
In: Snow, C.A. & Ferguson, Ch.A. (eds.) *Talking to Children. Language Input and Acquisition*, Cambridge: Cambridge University Press, Chapter 13, 279-308.

Bol, G. & Kuiken, F. (1988)
Grammaticale Analyse van Taalontwikkelingsstoornissen. Academisch Proefschrift ter verkrijging van de graad van doctor in de Letteren aan de Universiteit van Amsterdam, 21 juni 1988.

Bol, G. & Kuiken, F. (1989)
Handleiding GRAMAT. Een methode voor het diagnostiseren en kwalificeren van taalontwikkelingsstoornissen. Nijmegen: Berkhout.

Bowerman, M. (1973)
Early syntactic development. A Cross-linguistic study with special reference to Finnish. Cambridge: Cambridge University Press.

Bowerman, M. (1982)
Starting to Talk Worse: Clues to language acquisition from children's late speech errors.
In: Strauss, S. (ed.) *U-shaped behavioral growth*, New York: Academic Press, Chapter 5, 101-145.

Brinton, B. & Fujiki, M. (1984)
The development of topic manipulation skills in discourse.
Journal of Speech and Hearing Research, 27, 350-358.

Brinton, B.; Fujiki, M.; Loeb, D.F. & Winkler, E. (1986)
Development of Conversational Repair Strategies in Response to Requests for Clarification.
Journal of Speech and Hearing Research, 29, 75-81.

Brown, G. & Yule, G. (1983)
: *Discourse Analysis*. Cambridge Textbooks in Linguistics. Cambridge: Cambridge University Press.

Brown, R. (1973)
: *A First Language. The Early Stages*. Cambridge, Mass.: Harvard University Press.

Bruner, J.S. (1975)
: The ontogenesis of speech acts. *Journal of Child Language*, 2, 1-20.

Bruner, J.S. (1978)
: The Role of Dialogue in Language Acquisition. In: Sinclair, Jarvella & Levelt (eds.), *The Child's Conception of Language*. Heidelberg: Springer-Verlag Berlin, 241-256.

Bruner, J.S. (1983)
: *Child's Talk: Learning to use language*. Oxford: Oxford University Press.

Bullowa, M. (ed.) (1979)
: *Before speech: The beginning of interpersonal communication*. Cambridge: Cambridge University Press.

Capella, J.N. (1981)
: Mutual influence in expressive behaviour: Adult-adult and infant-adult interaction. *Psychological Bulletin*, 89, 101-132.

Capella, J.N. & Street, R.L. (eds.) (1985)
: A functional approach to the structure of communicative behaviour. In: Street, R.L. & Capella, J.N. (eds.), *Sequence and Pattern in Communicative Behaviour*. The Social Psychology of Language 3. London: Edward Arnold. Introduction, 1-29.

Chaika, E. & Lambe, R.A. (1989)
: Cohesion in schizophrenic narratives, Revisited. *Journal of Communication Disorders*, 22, 407-421.

Chapman, K.L.; Leonard, L.B. & Mervis, C.B. (1986)
: The effect of feedback on young children's inappropriate word usage. *Journal of Child Language*, 13, 101-117.

Chapman, R.S. (1981)
: Mother-child interaction in the second year of life: Its role in language development. In: Schiefelbusch, R.L. & Bricker, D.D. (eds.) *Early Language: Acquisition and Intervention*. Baltimore: University Park Press, 201-250.

Cherry, L.J. (1979)
: The role of Adult's Requests for Clarification in the Language Development of Children. In: Freedle, R. (ed.) *New Directions in Discourse Processing: A multidisciplinary approach*, Norwood, New York: Ablex, Volume 2, 273-286.

Chomsky, N. (1965)
: *Aspects of the theory of syntax*. Cambridge, Mass.: The MIT Press.

Clahsen, H. (1988)
: *Normale und gestörte Kindersprache. Linguistische Untersuchungen zum Erwerb von Syntax und Morphologie*. John Benjamins Publ. Cie: Amsterdam, Philadelphia.

Cohen, J. (1960)
A Coefficient of Agreement for Nominal Scales. *Educational and Psychological Measurement*, 20, 1, 37-46.

Colmar, S. (1987)
Review of G. Wells 'Language development in the pre-school years',
Cambridge: Cambridge University Press, 1985.
First Language, 7, 20, 173-175.

Conti-Ramsden, G. (1987)
Mother-Child talk with language impaired children. Paper presented at the first AFASIC symposium, University of Reading, 1987.

Conti-Ramsden, G. (1988)
Mothers in dialogue with language-impaired children.
Topics in Language disorders, 5, 58-68.

Conti-Ramsden, G. (1989)
Developmental Language Disorders.
In: Grundy, K. (ed.) *Linguistics in Clinical Practice*, London : Taylor & Francis, chapter 13, 242-254.

Conti-Ramsden, G. (1990)
Maternal recasts and other contingent replies to language-impaired children. *Journal of Speech and Hearing Disorders*, 55, 262-274.

Conti-Ramsden, G. & Friel-Patti, S. (1983)
Mother's discourse adjustments to language-impaired and non-language-impaired children.
Journal of Speech and Hearing Disorders, 48, 360-367.

Cook, Th. D. & Campbell, D.T. (1979)
Quasi-Experimention. Design & Analysis Issues for Field Settings.
Chicago: Rand McNally College Publ.Co.

Corrigan, R. (1980)
Use of Repetition to Facilitate Spontaneous Language Acquisition.
Journal of Psycholinguistic Research, 9, 3, 231-241.

Craig, H.G. (1983)
Applications of Pragmatic Language Models for Intervention.
In: Gallagher, T.M. & Prutting, C.A. (eds.) *Pragmatic Assessment and Intervention Issues in Language*, San Diego: College-Hill Press, 101-127.

Craig, R.T. & Tracy, K. (eds.) (1983)
Conversational Coherence: Form, Structure and Strategy. Beverly Hills: Sage Publications.

Cross, T.G. (1977)
Mother's speech adjustments: The contribution of selected child listener variables.
In: Snow, C. & Ferguson, Ch. A. (eds.) *Talking to children: Language input and acquisition*. Cambridge, UK: Cambridge University Press. Chapter 6, 151-188.

Cross, T.G. & Morris, J.E. (1980)
Linguistic feedback and maternal speech: Comparisons of mothers adressing infants, one-year-olds and two-year-olds.
First Language, 1, 2, 98-121.

Crystal, D. (1981)
Clinical Linguistics. Wenen: Springer Verlag.

Crystal, D.; Fletcher, P. & Garman, M. (1976)
The Grammatical Analysis of Language Disability: A procedure for assessment and remedi-

ation. London: Edward Arnold.
Dale, P.S. (1980)
   Is early pragmatic development measurable?
   Journal of Child Language, 7, 1-12.
Day, P.S. (1986)
   Deaf Children's Expression of Communicative Intentions.
   Journal of Communication Disorders, 19, 367-385.
Demaio, L.J. (1982)
   Conversational Turn-Taking: A Salient Dimension of Children's Language Learning.
   In: Lass, New York (ed.) *Speech and Language: Advances in Basic Research and Practice*, Volume 8. New York: Academic Press, 159-190.
Demetras, M.J.; Post, K.N. & Snow, C.E. (1986)
   Feedback to first language learners: The role of repetitions and clarification requests.
   Journal of Child Language, 13, 275-292.
Dijk, T.A. van (1977)
   Text and Context. Explorations in the semantics and pragmatics of discourse. London and New York: Longman Linguistic Library nr.21.
Dijk, T.A. van (1979)
   Pragmatic Connectives.
   Journal of Pragmatics, 3, 447-456.
Dijk, T.A. van (ed.) (1985)
   Handbook of Discourse Analysis. London: Routledge & Kegan Paul.
Dore, J. (1974)
   A Pragmatic Description of Early Language Development.
   Journal of Psycholinguistic Research, 3, 4, 343-350.
Dore, J. (1978)
   Variation in preschool children's conversational performances.
   In: Nelson, K.E. (ed.) *Children's Language*, Volume 1. New York: Gardner Press, Inc. Chapter 9, 397-444.
Dromi, E. (1989)
   The Significance of Input and Interaction in Children Learning Language under Exceptional Circumstances.
   Discussion Paper, presented at the Tenth Biennial Meeting of ISSBD, Jyvaskyla, Finland, July 1989.
Dromi. E. & Beny-Noked, S. (1984)
   Topic initiation and sustaining in conversations of Language Impaired Children with their Mothers. Paper presented at the Ninth Boston University Conference on Language Development, October 1984.
Dubber, C. & Weijdema, W. (1980)
   Functies van herhaling in verbale interactie.
   In: Foolen, A.; Hardeveld, J. & Springorum, D. (red.) *Conversatie Analyse*, KUN, Nijmegen, 224-241.
Duncan, S. Jr. & Fiske, D.W. (1977)
   Face-to-Face interaction. Hilldale, New York: Lawrence Earlbaum.
Duncan, S. Jr. & Fiske, D.W. (1985)
   The Turn System.
   In: Duncan, S Jr. & Fiske, D.W. (eds.) *Interaction Structure and Strategy.* Cambridge:

Cambridge University Press. Chapter 3, 43-64.
Dunn, L. (1976)
Expanded manual: Peabody Picture Vocabulary Test, American Guidance Service, Circle Pines, Minn.
Edmonson, W. (1981)
Spoken Discourse. A Model for Analysis. London: Longman.
Eisenson, J. (1972)
Aphasia in Children. New York: Harper & Row.
Evans, M.A. (1987)
Discourse characteristics of reticent children.
Applied Psycholinguistics, 8, 171-184.
Evans, M.A. (1989)
Verbal Interactions with Reticent Children. Paper presented at the Tenth Biennial Meeting of ISSBD, Jyvaskyla, Finland, July 1989.
Ervin-Tripp, S. (1979)
Children's Verbal Turn-Taking.
In: Ochs, E. & Schieffelin, B.B. (eds.) *Developmental Pragmatics*.
New York: Academic Press. Chapter 9, 391-429.
Ervin-Tripp, S. & Strage, A. (1986)
Parent-Child Discourse.
In: Van Dijk, T.A. (ed.) *Handbook of Discourse Analysis*. Volume 3, Discourse and Dialogue. London: Academic Press. Chapter 6, 67-77.
Ferguson, Ch.A. (1977)
Baby Talk as simplified Register.
In: Snow, C.A. & Ferguson, Ch.A. (eds.) *Talking to Children. Language input and acquisition*, Cambridge, UK: Cambridge University Press, Chapter 9, 219-235.
Fey, M. & Leonard, L. (1983)
Pragmatic Skills of Children with Specific Language Impairment.
In: Gallagher, T. & Prutting, C. (eds.) *Pragmatic Assessment and Intervention Issues in Language*, San Diego: College Hill Press, 65-82.
Fey, M.; Leonard, L. & Wilcox, K. (1981)
Speech-style modifications of language impaired children.
Journal of Speech and Hearing Disorders, 46, 91-97.
Fine, J. (1978)
Conversation, Cohesive and Thematic Patterning in Children's Dialogues.
Discourse Processes, 1, 247-266.
(Also in: Franklin, M.B. & Barter, S.S. (eds.) (1988) *Child Language. A Reader*, New York, Oxford: Oxford University Press, Chapter 21, 251-262).
Foster, S.H. (1986)
Learning discourse topic management in the preschool years.
Journal of Child Language, 13, 231-250.
Franco, F. & D'Odorico, L. (1988)
Baby Talk from the Perspective of Discourse Production: Linguistic Choices and the Context Coding by Different Speakers.
Journal of Psycholinguistic Research, 17, 1, 29-63.
Freedle, R.O. & Fine, J. (1983)
An Interactional Approach to the Development of Discourse.
In: Fine, J. & Freedle R.O. (eds.) *Developmental Issues in Discourse*, Norwood, New York: Ablex, Chapter 5, 143-168.

French, P. & MacLure, M. (eds.) (1981)
　Adult-Child Conversation. London: Croom Helm.
French, P. & Woll, B. (1981)
　Context, meaning and strategy in parent-child conversation.
　In: Wells, G. (ed.) *Learning through interaction*, Cambridge: Cambridge University Press. Chapter 4, 157-182.
Friemoth L.R. & Ashmore, L.L. (1983)
　Perceptive and expressive Wh-question performance by language delayed children. *Journal of Communication Disorders*, 16, 99-109.
Friel-Patti, S. & Conti-Ramsden, G. (1984)
　Discourse Development in Atypical Language Learners.
　In: Kuczaj II, S.A. (ed) *Discourse Development. Progress in Cognitive Development Research*. New York: Springer-Verlag, Chapter 8, 167-194.
Fuller-Fulero, L. (1983)
　Informational functions of mother-child discourse: Knowing when we see them. *Journal of Child Language*, 10, 223-229.
Furrow, D.; Nelson, K. & Benedict, H. (1979)
　Mothers' speech to children and syntactic development: Some simple relationships.
　Journal of Child Language, 6, 423-442.
Gallagher, T. (1977)
　Revision behaviours in the speech of normal children developing language.
　Journal of Speech and Hearing Research, 20, 303-318.
Gallagher, T. (1980)
　Contingent query sequences within adult-child discourse.
　Journal of Child Language, 8, 51-62.
Gallagher, T. & Darnton, B. (1978)
　Conversational aspects of the speech of language disordered children: Revision behaviours.
　Journal of Speech and Hearing Research, 21, 118-135.
Gallagher, T.M. & Prutting, C.A. (eds.) (1983)
　Pragmatic Assessment and Intervention Issues in Language. San Diego: College Hill Press.
Garfinkel, H. (1967)
　Studies of the routine grounds of everyday activities.
　Studies in Ethnomethodology, Englewood Cliffs, NJ: Prentice Hall, 35-76.
Garnham, A.; Oakhill, J. & Johnson-Laird, P.N. (1982)
　Referential continuity and the coherence of discourse.
　Cognition, 11, 29-46.
Garvey, C. & BenDebba, M. (1978)
　An experimental Investigation of Contingent Query Sequences.
　Discourse Processes, 1, 36-50.
Garvey, C. & Berninger, G. (1981)
　Timing and turn taking in children's conversation.
　Discourse Processes, 4, 562-568.
Van der Geest, T. (1977)
　Some interactional aspects of language acquisition.
　In: Snow, C. & Ferguson, Ch. A. (eds.) *Talking to Children. Language Input and Acqui-*

sition, Cambridge: Cambridge University Press, Chapter 4, 89-107.

Gelman, R. & Shatz, M. (1977)
Appropriate Speech Adjustments: The Operation of Conversational Constraints on Talk to Two-Year-Olds.
In: Lewis, M. & Rosenblum, L.A. (eds.) *Interaction, Conversation and the Development of Language*, New York: John Wiley & Sons, Chapter 2, 27-61.

Ginsburg & Opper (1979)
Piaget's Theory of Intellectual Development. Second edition.
Englewood Cliffs: Prentice Hall Inc.

Gleason, J.B. (1977)
Talking to Children: Some notes on feedback.
In: Snow, C. & Ferguson, Ch. A. (eds.) *Talking to Children. Language Input and Acquisition*, Cambridge: Cambridge University Press, Chapter 8, 199-205

Gleitman, L.R.; Newport, E.L. & Gleitman, H. (1984)
The current status of the motherese hypothesis.
Journal of Child Language, 11, 43-79.

Goldberg, J.A. (1983)
A move towards describing conversational coherence.
In: Craig, R.T & Tracy, K. (eds.) *Conversational Coherence: Form, Structure and Strategy*, Beverly Hills: Sage Publ., Chapter 1, 25-45.

Goodenough, D.R. & Weiner, S.L. (1978)
The Role of Conversational Passing Moves in the Management of Topical Transitions. *Discourse Processes*, 1, 395-404.

Grice, P. (1975)
Logic and Conversation.
In: Cole, P. & Morgan, J.L. (eds.) *Syntax and Semantics III; speech acts*, New York: Academic Press, 41-59.

Grosz, B.J. & Sidner, C.L. (1986)
Attention, Intentions, and the Structure of discourse.
Computational Linguistics, 12, 3, 175-204.

Grunwell, P. & James, A. (1989)
The functional Evaluation of Language Disorders. London, New York, Sydney: Croom Helm.

Guralnick, M.J. & Paul-Brown, D. (1986)
Communicative interactions of mildly delayed and normally developing preschool children: Effects of listener's developmental level.
Journal of Speech and Hearing Research, 29, 2-10.

Gurman Bard, E. & Anderson, A.H. (1983)
The Unintelligibility of speech to children.
Journal of Child Language, 10, 265-292.

Haft-Van Rees, M.A. (1987)
Van opeenvolgingsregels tot bijeenhorende paren: De samenhang tussen taaluitingen in gesprekken.
TTT, Interdisciplinair Tijdschrift voor Taal- & Tekstwetenschap, 7, 239-256.

Haft-Van Rees, M.A. (1989)
Taalgebruik in gesprekken. Inleiding in gespreksanalytisch onderzoek. Martinus Nijhoff, Leiden, 1989.

Halliday, M.A.K. (1975)
Learning how to mean. Explorations in the development of language. London: Edward Arnold.

Halliday, M.A.K. & Hasan, R. (1976)
Cohesion in English. London: Longman.
Hargrove, P.M.; Straka, E.M. & Medders, E.G. (1988)
Clarification requests of normal and language-impaired children.
British Journal of Disorders of Communication, 23, 51-62.
Hartveldt, R. (1987)
Pragmatic Aspects of Coherence in Discourse. Proefschrift ter verkrijging van het doctoraat aan de Rijksuniversiteit te Groningen, 25 november, 1987.
Haslett, B. (1987)
An Integrated Perspective on Human Communication.
In: Haslett, B. *Communication: Strategic Action in Context*, Hillsdale, New York: Lawrence Earlbaum Publ., Chapter 6, 115-147.
Have, P. Ten (1982)
Sociologische Gespreksanalyse.
TTT, Interdisciplinair Tijdschrift voor Taal- & Tekstwetenschap, 2, 2, 90-104.
Have, P. Ten (1987)
Transcriptie, interpretatie, analyse: De methodologie van gespreksanalyse.
In: Ten Have, P. *Sequenties en formuleringen*. Proefschrift ter verkrijging van de graad van doctor in de Letteren aan de Universiteit van Amsterdam, 1987, Hoofdstuk III.1, 292-325.
Hekken, S.M.J. van & Roelofsen, W. (1982)
More questions than answers: A study of question-answer sequences in a naturalistic setting.
Journal of Child Language, 9, 445-460.
Heim, M. (1989)
Kommunikatieve vaardigheden van niet of nauwelijks sprekende kinderen met infantiele encephalopathie. Een analyse van de kommunikatieve interaktie tussen niet-sprekende kinderen en hun dagelijkse konversatiepartners. Instituut voor Algemene Taalwetenschap, Universiteit van Amsterdam, Amsterdam.
Hickmann, M. (1987)
Social and functional approaches to language and thought.
Orlando: Academic Press, Inc.
Hirsch-Pasek, K. & Treiman, R. (1984)
Brown & Hanlon revisited: Mother's sensitivity to ungrammatical forms.
Journal of Child Language, 11, 81-88.
Hobbs, J.R. (1979)
Coherence and Coreference.
Cognitive Science, 3, 67-90.
Hobbs, J.R. (1983)
Why is Discourse Coherent.
In: Neubauer, Fr. (ed.) *Coherence in natural-language texts*, Hamburg: Helmut Buske Verlag, 29-70.
Hoff-Ginsberg, E. (1987)
Topic relations in mother-child conversation.
First Language, 7, 20, 145-158.
Hogan, A.E. & Seibert, J.M. (1984)
A Developmental Stage-Based Assessment Instrument for Early Communication Development.
In: Berg, J.M. (ed.) *Perspectives and Progress in Mental Retardation*, Volume 1: Social,

Psychological and educational Aspects. IASSMD edition.
Hopper, R. (1983)
Interpretation as Coherence Production.
In: Craig, R.T. & Tracy, K. (eds.) *Conversational Coherence: Form, Structure and Strategy.* Beverly Hills: Sage Publications, Chapter 4, 81-98.
Houtkamp-Steenstra, H. & Mazeland, H. (1982)
Beurten en grotere gesprekseenheden.
TTT, Interdisciplinair Tijdschrift voor Taal- & Tekstwetenschap, 2, 2, 105-135.
Hunt, K.W. (1970)
Syntactic maturity in school children and adults.
Monographs of the Society for Research in Child Development, serial 134, 35, 1.
Hurtig, R. (1977)
Toward a functional theory of discourse.
In: Freedle, R.D. (ed.) *Discourse Production and Comprehension*, Norwood, New York: Ablex.
Hustler, D. (1981)
Some Comments on Clarification Requests: A Response to Langford
In: French, P. & MacLure, M. (eds.) *Adult-Child Conversation*, London: Croom Helm, 175-184.
Hymes, D. (1974)
Foundations in Sociolinguistics: An ethnographic approach. Philadelphia: University of Pennsylvania Press.
Ierland, M.S. van (1982)
Normen, normering en taalontwikkelingsstoornissen.
TTT, Interdisciplinair Tijdschrift voor Taal- & Tekstwetenschap, 2, 199-212.
Ierland, M.S. van (1987)
1967-1987, Een kleine balans van twintig jaar psycholinguïstisch onderzoek naar Taalontwikkelingsstoornissen. Voordracht bij het afscheid van Ben Tervoort, 28 februari 1987, OPI Amsterdam.
Ingram, T.T.S. (1961)
Specific Developmental Disorders in Speech in Childhood.
Brain, 82, 450-467.
Jacobs, S. & Jackson, S. (1983)
Speech Act Structure in Conversation. Rational Aspects of Pragmatic Coherence.
In: Craig, R.T. & Tracy, K. (eds.)
Conversational Coherence: Form, Structure and Strategy. Beverly Hills: Sage Publications, Chapter 2, 47-65.
Jefferson, G. (1972)
Side Sequences.
In: Sudnow, D. (ed.) *Studies in Social Interaction,*
New York: Free Press, 294-338.
Johnston, J.R. (1986)
The Discourse Symptoms of Developmental Disorders.
In: Van Dijk, T.A. (ed.) *Handbook of Discourse Analysis.* Volume 3: Discourse and Dialogue, London: Academic Press, Chapter 7, 79-93.
Jose, P.E. (1988)
Sequentiality of Speech Acts in Conversational Structure.
Journal of Psycholinguistic Research, 17, 1, 65-88.

Junefelt, K. (1989)
Blindness of the mother or of the child and the development of communication. Paper presented at the Tenth Biennial Meeting of ISSBD, Jyvaskyla, Finland, July 1989.

Kamhi, A.G. (1989)
Language disorders in Children.
In: Leahey, M. (ed.) *Disorders of Communication: The Science of Intervention*. London: Taylor & Francis, chapter 6, 69-102.

Karmiloff-Smith, A. (1979)
A functional approach to child language: A study of determiners and reference. Cambridge, Mass.: Cambridge University Press.

Kay, D.A. & Anglin, J.M. (1982)
Overextension and underextension in the child's expressive and receptive speech.
Journal of Child Language, 9, 83-98.

Kaye, K. & Charney, R. (1980)
How mothers maintain "dialogue" with two-year-olds.
In: Olson, D. (ed.) *The Social foundations of language and thought*, New York: Norton.

Kaye, K. & Charney, R. (1981)
Conversational Asymmetry between mothers and children.
Journal of Child Language, 8, 35-49.

Keenan, E. & Klein, E. (1975)
Coherency in children's discourse.
Journal of Psycholinguistic Research, 4, 365-380.

Keenan, E. & Schieffelin, B. (1976)
Topic as a discourse notion.
In: Li, C. (ed.) *Subject and Topic*, New York: Academic Press, 337-384. (Also in: Ochs, E. & Schieffelin, B. (eds.) (1983) *Acquiring Conversational Competence*, London: Routledge & Kegan Paul, 337-384).

Kessel, F.S. (1988)
The Development of Language and Language Researchers. Essays in Honor of Roger Brown. Hillsdale, New York: Lawrence Earlbaum.

Klecan-Aker, J.S. & Swank, P.R. (1988)
The Use of a Pragmatic Protocol with Normal Preschool Children.
Journal of Communication Disorders, 21, 85-102.

Kleeck, A. van; Maxwell, M. & Gunter, Ch. (1985)
A methodological study of Illocutionary Coding in Adult-Child Interaction. *Journal of Pragmatics*, 9, 659-681.

Kreckel, M. (1981)
Communication acts and shared knowledge in natural discourse.
London: Academic Press.

Langford, D. (1981)
The clarification request sequence in conversation between mothers and their children.
In: French, P. & MacLure, M. (eds.) *Adult-Child Conversation*, London: Croom Helm, 159-174.

Lasky, E. & Klopp, K. (1982)
Parent-child interactions in normal and languagedisordered children.
Journal of Speech and Hearing Disorders, 47, 7-18.

Ledbetter, P.J. & Dent, C.H. (1988)
Young children's sensitivity to direct and indirect request structure.

First Language, 8, 24, 227-246.
Lee, L. (1966)
Development of sentence types: A method of comparing normal and deviant syntactic development.
Journal of Speech and Hearing Disorders, 31, 311-330.
Leonard, L.B. (1972)
What is deviant language ?
Journal of Speech and Hearing Disorders, 37, 427-446.
Leonard, L.B. (1986)
Conversational Replies of Children with Specific Language Impairment.
Journal of Speech and Hearing Research, 29, 114-119.
Leonard, L.B.; Bolders, J.G. & Miller, J.A. (1976)
An examination of the semantic relations reflected in the language usage of normal and language disordered children.
Journal of Speech and Hearing Research, 19, 371-392.
Levinson, S. (1983)
Pragmatics. Cambridge: Cambridge University Press.
Lewis, Ch. & Gregory, S. (1987)
Parents' talk to their infants: The importance of context.
First Language, 7, 201-216.
Lewis, M. & Rosenblum, L.A. (eds.) (1977)
Interaction, Conversation and the Development of Language.
New York: John Wiley & Sons.
Liles, B. (1985)
Cohesion in the narratives of normal language and disordered children.
Journal of Speech and Hearing Research, 28, 123-133.
Lotz Stine, E. & Bohannon III, J.N. (1983)
Imitations, interactions, and language acquisition.
Journal of Child Language, 10, 589-603.
Luszcz, M.A. & Bacharach, V.R. (1983)
The emergence of communicative competence: Detection of conversational topics.
Journal of Child Language, 10, 623-637.
Lyons, J. (1977)
Semantics. Cambridge: Cambridge University Press.
Maratsos, M. (1988)
Cross-linguistic Analysis, Universals, and Language Acquisition.
In: Kessel, F.S. (ed.) *The Development of Language and Language Researchers, Essays in Honor of Roger Brown*. Hillsdale, New York: Lawrence Earlbaum, Chapter 7, 121-152.
Martin, J.A.; Maccoby, E.E.; Baran, K.W. & Jacklin, C.N. (1981)
Sequential Analysis of mother-child interaction at 18 months:
A comparison of microanalytic methods.
Developmental Psychology, 17, 146-157.
Matarazzo, J.D. & Wiens, A.D. (1972)
The interview: Research on its anatomy and structure. Chicago: Aldine-Atherton.
Matlew, M. (1980)
Mothers' control strategies in dyadic mother/child conversations.
Journal of Psycholinguistic Research, 9, 327-347.

McDade, H.L. (1981)
A parent-child interactional model for assessing and remediating language disabilities. *British Journal of Disorders of Communication*, 16, 1, 175-183.

McDonald, L. & Pien, D. (1982)
Mother's conversational behaviour as a function of interactional content.
Journal of Child Language, 9, 337-358.

McNeill, D. (1966)
Developmental Psycholinguistics.
In: Smith, F. & Miller, G. (eds.) *The genesis of language: A psycholinguistic approach*, Cambridge, Mass: The MIT Press.

McReynolds, L.V. & Kearns, K.P. (1983)
Single-Subject Experimental Designs in Communicative Disorders.
Baltimore: University Park Press.

McTear, M.F. (1981)
Towards a Model for Analysing Conversation Involving Children.
In: French, P. & MacLure, M. (ed.) *Adult-Child Conversation*, London: Croom Helm, Chapter 11, 187-209.

McTear, M.F. (1985)
Children's Conversation. Oxford: Basil Blackwell Publ. Ltd.

McTear, M. & Conti-Ramsden (1989)
Assessment of Pragmatics.
In: Grundy, K. (ed.) *Linguistics in clinical practice*. London: Taylor & Francis. Chapter 8, 150-167.

Mentis, M. & Prutting, C.A. (1987)
Cohesion in the discourse of normal and head-injured adults.
Journal of Speech and Hearing Research, 30, 88-98.

Mishler, E.G. (1975)
Studies in dialogue and discourse: An exponential law of successive questioning.
Language and Society, 4, 31-51.

Morehead, D. & Ingram, D. (1973)
The development of base syntax in normal and linguistically deviant children.
Journal of Speech and Hearing Research, 16, 330-352.

Morgan, J.L. (1982)
Discourse Theory and the independance of sentence grammar.
In: Tanner. D. (ed.) *Analyzing Discourse: Text and Talk*. George Town Round Table on Language and Linguistics, 1981. George Town University Press 1982, 196-204.

Morris, G.H. & Hopper, R. (1980)
Remediation and legislation in everyday talk: How communicators achieve consensus and rules.
Quarterly Journal of Speech, 67, 266-274.

Nelson, K.; Denninger, M.; Bonvillian, J; Kaplan, B. & Baker, N. (1984)
Maternal input adjustments and non-adjustments as related to children's linguistic advances and to language acquisition theories.
In: Pelligrini, A. & Yawkey, T. (eds.) *The development of oral and written language in social contexts*. New York: Ablex.

Neubauer, F. (ed.) (1983)
Coherence in Natural-Language Texts. Hamburg: Helmut Buske Verlag.

Newport, E.L.; GLeitman, H. & Gleitman, L.R. (1977)
Mother, I'd rather do it myself: Some effects and non-effects of maternal speech style.

In: Snow, C.A. & Ferguson, Ch.A. (eds.) *Talking to Children. Language Input and Acquisition*, Cambridge: Cambridge University Press, Chapter 5, 109-149.

O'Brian, M. & Nagle, K.J. (1987)
Parents' speech to toddlers: The effect of play context.
Journal of Child Language, 14, 269-279.

Ochs, E. (1979)
Transcription as theory.
In: Ochs, E. & Schieffelin, B.B. (eds.) *Developmental Pragmatics*, New York: Academic Press.

Ochs Keenan, E. (1983)
Conversational Competence in Children.
Ochs, E. & Schieffelin, B.B. *Acquiring Conversational Competence*, London: Routledge & Kegan Paul, Chapter 1, 3-25.

Ochs Keenan, E. & Klein, E. (1975)
Coherency in Children's Discourse.
Journal of Psycholinguistic Research, 4, 4, 365-379.

Pearce, W.B. & Conklin, F. (1979)
A model of hierarchical meanings in coherent conversation and a study of 'indirect responses'.
Communication Monographs, 46, 75-87.

Penman, R.; Cross, T.; Milgrom-Friedman, J. & Meares, R. (1983)
Mothers' speech to prelingual infants: A pragmatic analysis.
Journal of Child Language, 10, 17-34.

Perner, J. & Leekam, S.R. (1986)
Belief and quantity: Three-year olds' adaptation to listener's knowledge.
Journal of Child Language, 13, 305-315.

Piaget, J. (1959)
The language and thought of the child. London: Routledge & Paul.

Platt, C.B. & MacWhinney, B. (1983)
Error assimilation as a mechanism in language learning.
Journal of Child Language, 10, 401-414.

Polanyi, L. (1985)
A theory of discourse structure and discourse coherence.
In: Heilfort, W.H.; Kroeber, P.D. & Peterson, K.L. (eds.), *CLS 21, Part 1: Papers from the General Session at the Twenty-First Regional Meeting, Chicago Linguistic Society*, 306-322.

Prather, E.; Cromwell, K. & Kenney, K. (1989)
Types of Repairs Used by Normally Developing and Language- Impaired Preschool Children in Response to Clarification Requests.
Journal of Communication Disorders, 22, 49-64.

Prinz, P.M. & Ferrier, L.J. (1983)
'Can you give me that one ?': The comprehension, production and judgment of directives in language impaired children.
Journal of Speech and Hearing Disorders, 48, 44-54.

Proctor, A. (1985)
Social Interaction as Related to Language Acquisition: An Annotated Bibliography.
Folia Phoniatrica, 37, 87-105.

Prutting, C.A. & Kirchner, D.M. (1987)
A Clinical Appraisal of the pragmatic aspects of language.

Journal of Speech and Hearing Disorders, 52, 105-119.

Reichman, R. (1978)
Conversational Coherency.
Cognitive Science, 2, 283-327.

Rees, N. (1973)
Auditory processing factors in language disorders: The view of Procustes' bed.
Journal of Speech and Hearing Disorders, 38, 304-315.

Reynell, J.K. (1974)
Manual for the Reynell Developmental Language Scales (revised),
The NFER, Nelson Publishing Company Ltd., Windsor.

Rieke, J.A.; Lynch, L.L. & Soltman, S.F. (1977)
Teaching Strategies for Language Development. New York: Grune & Stratton.

Robinson, R.J. (1987)
The Causes of Language Disorder: Introduction and Overview.
Proceedings of The First AFASIC Symposium, University of Reading, UK, 1987, 1-19.

Roth, F.P. & Spekman, N.J. (1984)
Assessing the pragmatic abilities of children: Part 1, Organizational Framework and assessment parameters.
Journal of Speech and Hearing Disorders, 49, 2-11.

Rumelhart, D. (1975)
Notes on a schema for stories.
In: Bobrow, D.D. & Collins, A. (eds.) *Representation and understanding*, New York: Academic Press, Studies in cognitive science.

Russo, J.B. & Owens, R.E. (1982)
The Development of an Objective Observation Tool for Parent-Child Interaction.
Journal of Speech and Hearing Disorders, 47, 165-173.

Sackett, G.P. (1979)
The lag-sequential analysis of contingency and cyclicity in behavioral interaction research.
In: Osofsky, J. (ed.) *Handbook of infant development*, New York: John Wiley & Sons, Chapter 17, 623-649.

Sacks, H.; Schegloff, E.A. & Jefferson, G. (1974)
A simplest systematics for the organization of turn-taking for conversation.
Language, 50, 4, 696-735. (also in: Schenkein, J. (ed.), 1978, *Studies in the organization of conversational interaction*, New York: Academic Press).

Sander. L.W. (1977)
The Regulation of Exchange in the Infant-Caretaker System and Some Aspects of the Context-Content Relationship.
In: Lewis, M. & Rosenblum, L.A. (eds.) *Interaction, Conversation and the Development of Language*, New York: John Wiley & Sons, chapter 6, 133-156.

Sanders, R.E. (1983)
Tools for Cohering Discourse and Their Strategic Utilization. Markers of Structural Corrections and Meaning Relations.
In: Craig, R.T. & Tracy, K. (eds.), *Conversational Coherence: Form, Structure and Strategy*, Beverly Hills: Sage Publications, Chapter 3, 67-80.

Schachter, F.F. (1979)
Everyday Mother Talk to Toddlers. Early Intervention.
New York: Academic Press.

Schaffer, H.R.; Hepburn, A. & Collis, G.M. (1983)
Verbal and nonverbal aspects of mothers' directives.
Journal of Child Language, 10, 337-355.

Schasfoort, M. (1984)
De konversatie-analytische beschrijving van gesprekken als sociale interaktie.
TTT, Interdisciplinair Tijdschrift voor Taal- & Tekstwetenschap, 4, 1, 41-67.

Schegloff, E.A. (1972)
Notes on a conversational practice: Formulating place.
In: Sudnow, D. (ed.) *Studies in social interaction*, New York: Free Press, 75-119.

Schegloff, E.A. & Sacks, H. (1973)
Openings up closings.
Semiotica, 7, 4, 289-327.

Schiffrin, D. (1987)
Why analyze discourse markers?
Discourse Markers, Cambridge: Cambridge University Press, 49-72.

Schwabe, A.M.; Olswang, L.B. & Kriegsman, E. (1986)
Requests for Information: Linguistic, Cognitive, Pragmatic, and Environmental Variables.
Language, Speech and Hearing Sciences in Schools, 38-55.

Schneiderman, M.H. (1983)
'Do what I mean, not what I say !' Changes in mothers' action directives to young children.
Journal of Child Language, 10, 357-367.

Schodorf, J.K. & Edwards, H.T. (1983)
Comparative analysis of parent-child interactions with language disordered and linguistically normal children.
Journal of Communication Disorders, 16, 71-83.

Schwartz, R.G.; Chapman, K.; Terrell, B.Y.; Prelock, P. & Rowan, L. (1985)
Facilitating word combination in language-impaired children through discourse structure.
Journal of Speech and Hearing Disorders, 50, 31-39.

Searle, J.R. (1969)
Speech Acts. Cambridge: Cambridge University Press.

Searle, J.R. (1975)
Indirect Speech Acts
In: Cole, P. & Morgan, J.L. (eds.) *Syntax and Semantics III: Speech Acts*, New York: Academic Press, 59-83.

Seibert, J.M. & Hogan, A.E. (1982)
Procedures manual for the Early Social-Communication Scales (ESCS). Unpublished manuscript. University of Miami Mailman Center for Child Development, Miami, Florida.

Seibert, J.M.; Hogan, A.E. & Mundy, P.C. (1984)
Developmental assessment of social-communication skills for early intervention: Testing a cognitive stage model.
In: Glow, R.A. (ed.) *Advances in the Behavioral Measurement of Children*, JAI Press, Greenwich, Connecticut.

Shatz, M. (1982)
On mechanisms of language acquisition: Can features of the communicative envi-

ronment account for development ?
In: Wanner, E. & Gleitman, L.R. (eds.) *Language Acquisition: The State of the Art*, Cambridge: Cambridge University Press, 102-127.

Shatz, M. & Gelman, R. (1977)
Beyond syntax: The influence of conversational constraints on speech modifications.
In: Snow, C.A. & Ferguson, Ch.A. (eds.) *Talking to Children: Language input and acquisition*, Cambridge, UK: Cambridge University Press, Chapter 7, 189-198.

Sigman, S.J. (1983)
Some considerations of the multiple constraints placed on conversational topics.
In: Craig, R.T. & Tracy, K. (eds.) *Conversational Coherence: Form, Structure and Strategy*, Beverly Hills, CA: Sage Publ.

Sinclair, J.McH. (1985)
On the Integration of Linguistic Description.
In: Van Dijk, T.A. (ed.) *Handbook of Discourse Analysis*, Volume 2; Dimension of Discourse, London: Academic Press, Chapter 2, 13-28.

Slobin, D.I. (1982)
Universal and particular in the acquisition of language.
In: Wanner, E. & Gleitman, L.R. (eds.) *Language Acquisition: The State of the Art*, Cambridge: Cambridge University Press, Chapter 5, 128-170.

Smith, C.B.; Adamson, L.B. & Bakeman, R. (1988)
Interactional Predictors of Early Language.
First Language, 8, 23, 143-156.

Snow, C.A. (1972)
Mothers' speech to children learning language.
Child Language, 43, 549-565.

Snow, C.A. (1986)
Conversations with Children.
In: Fletcher, P. & Garman, M. (eds.) *Language Acquisition*, Cambridge University Press, Cambridge, 1986, 69-89.

Snow, C.A. & Ferguson, Ch.A. (eds.) (1978)
Talking to Children: Language input and acquisition.
Cambridge, UK: Cambridge University Press.

Springorum, D. (1980)
Directieven, interpreterende reacties en conversationele voortzettingsmogelijkheden.
In: Foolen, A.; Hardeveld, J. & Springorum, D. (red.) *Conversatie Analyse*, KUN Nijmegen, 193-223.

Stark, R.E. & Tallal, P. (1981)
Selection of Children with Specific Language Deficits.
Journal of Speech and Hearing Research, 46, 2, 114-122.

Stella-Prorok, E.M. (1983)
Language development in the natural environment: A functional analysis of mother-child speech.
In: Nelson, K. (ed.) *Children's Language*, Volume 4, Hillsdale, New York: Earlbaum, 187-230.

Street, R.L. & Capella, J.N. (eds.) (1985)
Sequence and Pattern in Communicative Behaviour. The Social Psychology of Language 3. London: Edward Arnold.

Taylor, T.J. & Cameron, D. (1987)
  Functionalism and Exchange Structure in Conversation Analysis.
  In: Taylor, T.J. & Cameron, D. (eds.) *Analysing Conversation, Rules and Units in the Structure of Talk*, Oxford: Pergamon Press, Chapter 4, 65-80.
Tracy, K. (1985)
  Conversational Coherence: A cognitively grounded rules approach.
  In: Street, R.L. & Capella, J.N. (eds.), *Sequence and Pattern in Communicative Behaviour. The Social Psychology of Language 3*, London: Edward Arnold, Chapter 2, 30-49.
Vuchinich, S. (1977)
  Elements of cohesion between turns in ordinary conversation.
  Semiotica, 20, 229-257.
Wanner, E. & Gleitman, L. (eds.) (1982)
  Language Acquisition. The State of the Art. Cambridge: Cambridge University Press.
Wanska, S.K. & Bedrosian, J.L. (1985)
  Conversational Structure and Topic Performance in mother-child interaction.
  Journal of Speech and Hearing Research, 28, 579-584.
Warden, D. (1981)
  Learning to identify referents.
  British Journal of Psychology, 72, 93-99.
Watson, L. (1977)
  Conversational Participation by Language Deficient and Normal Children. ASHA 1977.
Wells, C.G. (1975)
  Coding Manual for the Description of Child Speech in its Conversational Context. University of Bristol School of Education, Revised Edition.
Wells, G. (ed.) (1981)
  Learning through interaction. The study of language development. Language at home and at school 1. Cambridge: Cambridge University Press.
Wells, G. (1985)
  Language Development in the Preschool Years. Language at home and at school 2. Cambridge: Cambridge University Press.
Wells, G.; MacLure, M. & Montgomery, M. (1981)
  Some strategies for sustaining conversation.
  In: Werth, P. (ed.) *Conversation and Discourse: Structure and Interpretation*, London: Croom Helm, Chapter 3, 73-85.
Wells, G.; Montgomery, M. & MacLure, M. (1979)
  Adult-Child Discourse: Outline of a Model of Analysis.
  Journal of Pragmatics, 3, 337-380.
Weijdema, W.; Dik, S.; Oehlen, M.; Dubber, C. & De Blauw, A. (1982)
  Structuren in verbale interactie. Strategieën van sprekers en hoorders in het taalgebruik. Muiderberg: Dick Coutinho.
Wulbert, M.; Inglis, S.; Kriegsmann, E. & Mills, B. (1975)
  Language Delay and Associated Mother-Child Interactions.
  Developmental Psychology, 11, 1, 61-70.

APPENDIX A

Transcript Conventions in IACV

This Appendix gives an account of the transcript conventions in IACV and is a summary of the text in 'The IACV Manual for Transcription and Analysis' (Van Balkom et al.,1990).

The transcript conventions are illustrated with examples. Extensive examples of transcripts are given in Appendix B.

APPENDIX A*
IACV Transcript Conventions

1. Verbal and nonverbal acts of both partners (adult and child) are transcribed subsequently in the order of occurrence during conversation.

   Verbal acts are segmented according to the definition of 'T-units', according to Hunt (1970). See also section 4.2.2.
   Verbal acts are transcribed in traditional orthography.

   Nonverbal Acts are segmented in singular, main activities. See also the segmentation-rules formulated in section 4.2.2.
   Nonverbal acts are circumscribed, indicating the main activity (e.g., "points to doll", "looks at A", "gives doll to A").

2. Adult/caregiver acts are indicated with an 'A' for adult/caregiver; the child's acts are indicated with a 'C' for child.

3. Context and situational information (and cues for helping to understand the content/intention of an act) are described between '<...>'. Contextual and situational information can be added to each transcribed act.

4. Nonfluencies, such as 'False Starts', 'Self-corrections' and 'Word/Word-part Repetitions' are put between two periods.
   The contextual information specifies the kind of nonfluency.

   Examples of Nonfluency
   .I. you are going?
   <False Start>

   I saw the .horse. cow in the box
   <Self-correction>

5. Unintelligible verbal acts or parts of verbal acts are transcribed as accurately as possible and put between (...).
   When it is impossible to guess the intended content, the number of syllables has to be put between the brackets.

---

* This appendix is an abstract of chapter 2 in 'The IACV Manual for Transcription and Analysis' (Van Balkom et al., 1990).

Examples of an Unintelligible verbal act
- I got (doll) from her
- (4 syllables)
  <difficult to understand because of whispering>

6. Unfinished verbal acts are transcribed until the break-off point. The break-off point is indicated by an '-'.

   Example of an Unfinished verbal act
   - Ivo, you just said that-
   - try to say, that is an-

7. Capitals are only used for verbal acts or parts of verbal acts which are produced with extra emphasis, such as exclamations, places/cities and in Christian names.

   Examples of using capitals
   - Truus, Jan, Kees, my aunt Fryda
   - Amsterdam, Hoensbroek
   - OKAY, TAKE care

8. Vocalisations are put in capitals and between double brackets.

   Examples of vocalisation
   - ((LAUGHING))
   - ((CRYING))
   - ((COUGHING))

   If the partners are singing, the text sung has to be interpreted as *one* act. The text or title of the song is put in the context; the transcript text refers to the act of singing by ((SINGING)).

   Example
   <A sings a song>
   A. ((SINGING))
       <happy birthday to you..>

9. The rise of intonation at the end of a verbal act is indicated by an '?'.

   Example of Rise of Intonation
   - where are you going to?
   - to school?

10. Extended pronunciation of a preceding phoneme is indicated by an ':'. (each ':' indicates a one second duration of the preceding phoneme).

   Examples of Extended pronunciation
   - I go to schoo:l
   - I'm CO::MING

11. Pauzes between consecutive verbal acts are indicated in the transcript immediately before the text of the verbal act. Pauses are indicated in digits.

12. Overlap in relations between consecutive verbal or nonverbal acts of one or both partners are indicated by '//' (start of the overlap) and ']' (end of the overlap).

   Examples of Overlap in relations
   C. //...
   A. ....]
      complete overlap of both partners

   C. //...
   A. ...]....
      complete to partial overlap of both partners

   C. ...//...
   A. ...]...
      partial to partial overlap of both partners.

   C. //...*...
   A. ...]
      the part //...* is in overlap with ...]

13. The transcripts of each session (based on fragments of 5 minutes) are divided into segments of 30 seconds. Each transcript consists of 10 segments of 30 seconds. The segments are indicated by a demarcation line between the brackets of the context area in the transcripts:
   <***>'

Note to the English reader
On the next pages examples of transcript segments of conversations between SLI and NLA children and their caregivers are given. As distinct from the transcript segments used in the examples in the previous chapters, we did not translate the transcripts in this appendix into English. The transcript segments in appendix B function only as illustrations of the way in which the transcription is performed and the computer is used.

APPENDIX B

IACV Transcript Example

In this Appendix four transcript segments are given as an illustration of the transcript procedures. The IACV transcripts are the result of a specific computer programme written for the purpose of this study. In addition to this computer programme for transcription two other programmes were written for the IACV Morpho-syntax Analysis and for the IACV Pragmatics Analysis.

The transcript segments are given in Dutch.

Dyad Number 9.: Sebastian (SLI-child) : Fragments from the sessions 2 and 9

Dyad Number 13: Rachel (NLA-Child) : Fragments from the sessions 2 and 9

KEY

Rec : Record nummer (volgorde nummer van acts)
<...> : context
(...) : paraphrases
A : Adult (v: verzorger)
C : Child (k: kind)
Nv : Nonverbal act code
T : Turn (beurt nummer)
I : Interval time (interval-tijd tussen opeenvolgende acts)

TRANSCRIPT OF RACHEL AND HER MOTHER (NLA DYAD)

Filename :   rachel2.tra
Last update: 8-10-1990
Date printed: 8 Oct 90 - 10:04:36
Adult: agnes,   Transcribed by: tiny

Rec	A/C Nv	T	V	I	Text of Acts Filename =rachel2.tra
1	C	1	1	0	//ehm <v. en k. zitten tegenover elkaar op stoeltjes tussen hen in op vloer staat kassa>
2	CA	1	1	0	bekijkt voorw. in haar hand] <(badje met popje)/ v. kijkt k.>
3	C	1	2	2	eh //enne lululu enne bo enne/x ko:pje <k. kijkt soms even op naar v.>
4	CB	1	2	0	wijst naar voorwerp] <>
5	A	1	1	0	o:ch een hoofd <>
6	C	2	3	0	een hoofdje eh //een- <afgebroken door v.>
7	A	2	2	0	of d'r kop maar nit 't kopje <(of de kop maar niet het kopje)>
8	C	3	4	0	is kopje <k. kijkt voorw./ v. kijkt k.>
9	A	3	3	1	nee <fluisterend>
10	C	4	5	0	//nee //kopje is daar <>
11	CA	4	5	0	kijkt vloer] <>
12	CB	4	5	0	wijst met voetje een kopje aan] <en kijkt naar v.>
13	A	4	4	0	ja en wat moot 'r nog mie han mevrouw? <(ja en wat moet u nog meer hebben mevrouw?)>
14	A	4	5	0	//d'r mot wal e bietje opschieten want hei kome nog m'eer mensen in de winkel <(u moet wel een beetje ...) (hei=hier) v.>

Rec	A/C Nv	T	V	I	Text of Acts	Filename =rachel2.tra
15	AA	4	5	0	//kijkt k.] <>	
16	CA	5	5	0	kijkt v.] <met open mond>	
17	A	5	6	2	ha:gelsjlag wat mot ur nog mie han? <(hagelslag, wat moet u nog meer hebben?)>	
18	C	6	6	0	ehm <k. kijkt vloer (denkt na)/ v. kijkt k.>	
19	A	6	7	0	eitjes? <>	
20	C	7	7	0	ja <>	
21	A	7	8	0	//eitjes <>	
22	AA	7	8	0	schrijft "eitjes" met vinger op rok] <k. kijkt voorw. in haar hand>	
23	A	7	9	0	en brood? <>	
24	C	8	8	0	eja <(ja)>	
25	A	8	10	0	//brood //ja <>	
26	AA	8	10	0	"schrijft" "brood"] <nonverebaal/k. kijkt activ. v.>	
27	AB	8	10	0	//kijkt k.] <>	
28	A	8	11	0	w'as dat 't mevrouw? <>	
29	C	9	9	0	ja <>	
30	A	9	12	0	//goed <>	
31	CA	10	9	0	steekt handje uit naar v.] <>	
32	AA	10	12	0	steekt hand uit naar k. <>	
33	CA	11	9	0	//tikt hand v. aan <>	

Rec	A/C Nv	T	V	I	Text of Acts	Filename =rachel2.tra
34	C	11	10	0	dank je wel] <>	
35	A	11	13	0	//kom mer hei mevrouw <(kom maar hier mevrouw)>	
36	AA	11	13	0	steekt andere hand uit naar k.] <>	
37	A	11	14	0	doch mer de had op <(doe maar de hand open)>	
38	CA	12	10	0	draait handpalm naar boven <>	
39	A	12	15	0	dat is //vlees* //sjinkewoe:sj* ehm: hagelsla:g* //eitjes <zelfverbetering/ (dat is vlees, hamworst,...)/k. kijkt op naar v.>	

TRANSCRIPT OF RACHEL AND HER MOTHER (NLA DYAD)

Filename: *rachel9.tra*
Last update: 8-10-1990
Date printed: 8 Oct 90 - 10:04:54
Adult: agnes,   Transcribed by: tiny

Rec	A/C Nv	T	V	I	Text of Acts Filename =rachel9.tra
1	AA	1	0	0	//roert in pannetje <v. en k. zitten op knieen bij fornuisje/ k. zit met rug naar camera gekeerd>
2	A	1	1	2	man wat smullen wij vandaag] <(man=tjonge)>
3	C	1	1	0	ja, <>
4	CA	1	1	0	speelt met potjes op fornuisje <niet goed te zien/ k. zit met rug naar camera>
5	A	2	2	0	weet je wat Judith kreeg bij tante Yvonne? <>
6	CA	2	1	0	//kijkt op naar v. <>
7	C	2	2	0	nee] <>
8	A	3	3	0	raad 's <v. roert verder/ k. speelt verder>
9	C	3	3	0	he, <>
10	A	4	4	0	wat maakt tante Yvonne 't vrijdags? <('t vrijdags= 'svrijdags)>
11	C	4	4	0	ehm //frieten <>
12	CA	4	4	0	kijkt v.] <>
13	AA	5	4	0	kijkt k.] <>
14	A	5	5	0	ho:y frietjes zegt ze <>
15	C	5	5	1	//wat zei Judith? <>

Rec	A/C Nv	T	V	I	Text of Acts	Filename =rachel9.tra
16	CA	5	5	0	//zit voorovergebogen over fornuisje en speelt met potjes] <>	
17	AA	6	5	0	kijkt k.] <>	
18	A	6	6	0	frietjes ho:: zegt ze <>	
19	A	6	7	0	zo (nu) maar even wachten <(of: doe)>	
20	C	6	6	0	zo, <>	
21	A	7	8	0	//kunnen we een kopje koffie gaan drinken he mevrouw? <>	
22	AA	7	8	0	//gaat zitten] <>	
23	CA	7	6	0	buigt over kist en //kijkt erin] <>	
24	C	7	7	0	hei::] <>	
25	AA	8	8	0	verschikt kopjes en schoteltjes op vloer <>	
26	CA	8	7	0	//pakt lepel uit kist en loopt ermee naar fornuis <>	
27	C	8	8	0	daar prik ik alles mee 'op] <>	
28	A	9	9	0	hm? <v. is nog bezig met kopjes en schoteltjes>	
29	C	9	9	0	prik ik alles mee op <>	
30	A	10	10	0	prik ik alles mee op? <(v. "begrijpt" k. niet)>	
31	C	10	10	0	//ja zo <>	
32	CA	10	10	0	maakt "schep-beweging" in de lucht] <>	

Rec	A/C Nv	T	V	I	Text of Acts Filename =rachel9.tra
33	A	11	11	0	ja dan prik je 'in de aardappeltjes en //de bloemkool <v. is bezig/ heeft niet gezien dat k. geen vork maar een lepel gebruikt>
34	C	11	11	0	//waar is de] aardappels? <>
35	CA	11	11	0	gaat voor fornuisje zitten] <>
36	A	12	12	0	//h'ier zaten de aardappeltjes in <v. kijkt k.>
37	AA	12	12	0	haalt dekseltje van keteltje] <>
38	CA	12	11	0	//zet lepel in de "aardappeltjes" <>
39	A	13	13	0	of ze gaar zijn he,] <>
40	C	13	12	0	//nee pak eens even je //mes <>
41	CA	13	12	0	draait van fornuis weg en kijkt richting "tafel"] <k. wil eten uitscheppen>
42	A	14	14	0	en] is dit op //prikken? <>

TRANSCRIPT OF SEBASTIAN AND HIS MOTHER (SLI DYAD)

Filename : sebas2.tra
Last update: 8-10-1990
Date printed: 8 Oct 90 - 10:05:13
Adult: moeder, Transcribed by: loetje

Rec	A/C Nv	T	V	I	Text of Acts Filename = sebas2.tra
1	CA	1	0	0	//wijst <v. en k. spelen met dierentuin>
2	C	1	1	0	dat?] <>
3	A	1	1	0	//dat? <>
4	AA	1	1	0	pakt olifant] <en zet het op de grond>
5	A	1	2	0	een olifantje <>
6	C	2	2	2	//(2 syll) <slecht gearticuleerd>
7	CA	2	2	0	pakt olifant in handen] <en kijkt ernaar>
8	A	2	3	0	roets nu toch <>
9	CA	3	2	0	//laat olifant van glijbaan glijden <>
10	C	3	3	2	oe: //oe] <begeleidend geluid>
11	A	3	4	0	oe:::] <olifant is beneden>
12	CA	4	3	0	pakt olifant en zet hem voor dierentuin <olifant valt om en stoot fles om>
13	AA	4	4	0	//zet dier op dak dierentuin <>
14	A	4	5	0	och] <reactie op omvallen olifant>
15	A	4	6	0	//moet 'olifantje dr'inken <>

Rec	A/C Nv	T	V	I	Text of Acts	Filename = sebas2.tra
16	AB	4	6	0	//pakt fles] <ligt op vloer naast olifant>	
17	CA	5	3	0	speelt even met dier op dak] <>	
18	A	5	7	0	kijk eens wil 't olifantje misschien toch drinken <dialect>	
19	A	5	8	0	//je kunt hem de fles geven <dialect>	
20	AA	5	8	0	houdt fles op naar k.] <>	
21	CA	6	3	0	//pakt fles <>	
22	A	6	9	0	het is een baby-olifantje] <>	
23	CA	7	3	0	//kijkt zoekend rond <>	
24	AA	7	9	0	stopt gedachteloos dier in wagen] <>	
25	A	7	10	0	//oh kleine baby kan niet 'eten zelf <>	
26	AB	7	10	0	pakt olifant en houdt hem voor k.] <>	
27	A	7	11	0	//geef hem maar de fl'es <>	
28	CA	8	3	0	pakt olifant en geeft hem de fles] <>	
29	C	8	4	0	//nee <>	
30	CB	8	4	0	//schudt hoofd van nee] <>	
31	CC	8	4	0	houdt olifant voor zich] <met gestrekte arm>	
32	A	8	12	0	wil hij niet //ja meh <>	
33	AA	8	12	0	//pakt olifant en zet hem op de grond] <>	
34	AB	8	12	0	kijkt naar k.] <>	

Rec	A/C Nv	T	V	I	Text of Acts	Filename = sebas2.tra
35	C	9	5	0	//datte ook <>	
36	CA	9	5	0	pakt ander dier en houdt het bij de fles] <>	
37	A	9	13	0	ook 't flesje? <>	
38	CA	10	5	0	//legt dier weer weg <>	
39	C	10	6	0	nee] <>	
40	A	10	14	0	ook niet <>	
41	CA	11	6	0	//pakt dier van dak en geeft het de fles <v. kijkt toe>	
42	C	11	7	0	dat ook] <>	
43	A	11	15	0	allemaal babydiertjes <>	
44	C	12	8	0	//nee: <>	
45	CA	12	8	0	legt dier terug op dak] <>	

Filename: *sebas9.tra*
Last update: 8-10-1990
Date printed: 8 Oct 90 - 10:05:30
Adult: moeder,   Transcribed by: loetje

Rec	A/C Nv	T	V	I	Text of Acts	Filename = sebas2.tra
1	AA	1	0	0	rijdt auto met popje erin over dierentuin <v. en k. zitten tegenover elkaar en spelen met dierentuin>	
2	AB	1	0	0	laat auto los <bovenaan glijbaan>	
3	CA	1	0	0	pakt auto vast en geeft hem een duwtje <auto rijdt van glijbaan af>	
4	CB	1	0	0	//kijkt naar auto op grond <>	
5	C	1	1	0	n'eet huilen] <eerste geluid op band (eerste 8 sec. geen geluid)>	
6	AA	2	0	0	//kijkt naar k. <>	
7	A	2	1	0	is niet aan 't huilen> <>	
8	A	2	2	0	oh] <begrijpend>	
9	CA	2	1	0	//wijst naar garage van huisje <schuin voor zich>	
10	C	2	2	0	nog een keer> <>	
11	C	2	3	0	//dat auto] <>	
12	AA	3	2	0	kijkt in wijsrichting] <>	
13	AB	3	2	0	pakt auto uit garage <>	
14	AC	3	2	0	//geeft auto aan k. <>	
15	A	3	3	0	die is te groot voor de (1 syll)> <>	

Rec	A/C Nv	T	V	I	Text of Acts	Filename = sebas2.tra
16	A	3	4	0	//die past er niet op jong] <>	
17	CA	3	3	0	bekijkt auto in handen] <en probeert popje erin te zetten>	
18	A	4	5	0	moet je een kleine auto pakken <>	
19	CA	4	3	0	pakt voorw. van vloer naast zich <popje>	
20	CB	4	3	0	//probeert popje in auto te zetten <>	
21	C	4	4	0	dat] <>	
22	A	5	6	0	nee dat past hei neet op <dialect>	
23	AA	5	6	0	wijst op dak dierentuin <op richel waar auto op moet rijden>	
24	A	5	7	0	//dat is te breed <>	
25	CA	5	4	0	kijkt in wijsrichting] <>	
26	AA	6	7	0	//kijkt naar k. <>	
27	A	6	8	0	past neet] <>	
28	CA	6	4	0	//kijkt om zich heen <>	
29	C	6	5	0	o:h] <>	
30	CB	6	5	0	//pakt voorw. van grond <met auto nog in hand>	
31	C	6	6	0	dat,] <>	
32	A	7	9	0	//kiek <>	
33	AA	7	9	0	pakt auto uit handen k.] <>	
34	A	7	10	0	(5 syll.) <>	

Rec	A/C Nv	T	V	I	Text of Acts	Filename = sebas2.tra
35	AB	7	10	0	//zet auto op dierentuin en rijdt ermee <op gedeelte waar hij wel past>	
36	A	7	11	0	//kijk zie je daar kan hij wel op rijden] <>	
37	CA	7	6	0	kijkt naar bewegingen v.] <>	
38	AA	8	11	0	laat auto los boven aan glijbaan <auto rijdt van glijbaan af>	
39	A	8	12	0	//oh goed zo <reactie op rijden van auto>	
40	AB	8	12	0	//kijkt naar auto] <>	
41	CA	8	6	0	kijkt naar auto] <>	
42	CB	8	6	0	pakt auto op <>	
43	CC	8	6	0	//wijst naar andere gedeelte van dierentuin <waar de grote auto niet past>	
44	C	8	7	0	en daar dan?] <>	

APPENDIX C

IACV Communication Functions

This Appendix contains definitions and examples of the Communication Functions used as coding categories and subcategories in the IACV Pragmatics Analysis Programme.

The text given is a summary of the 'The IACV Manual for Transcription and Analysis' (Van Balkom et al., 1990).

APPENDIX C
IACV Communication Functions*

CONTENTS

1. **Subsequence or Conversational Theme**

2. **Communication Functions of Subsequences**

3. **Control Functions**
3.1. General Request
3.2. Request for Justification
3.3. Request for Permission
3.4. Other Request
3.5. Wanting
3.6. Intend
3.7. Prohibition
3.8. Justification
3.9. Assent
3.10. Acknowledgement
3.11. Reject/Refuse
3.12. Other Controls

4. **Expressive Functions**
4.1. Exclamation
4.2. Accompaniment to Action
4.3. Query State or Attitude
4.4. Express State or Attitude
4.5. Encourage
4.6. Approval
4.7. Disapproval
4.8. Responses
4.9. Other Expressives

5. **Representational Functions**
5.1. Ostension
5.2. Statement
5.3. Content Question

---

* According to Wells (1975, 1985)

5.4. Response to Content Question
5.5. Yes/no Question
5.6. Affirm
5.7. Deny/Reject
5.8. Other Question
5.9. Other Responses
5.10. Justification
5.11. Other Representationals

6. **Tutorial Functions**
6.1. Demand for Required Form
6.2. Supply of Required Form
6.3. Question with Frame
6.4. Filler to Question with Frame
6.5. Question with Known Answer
6.6. Response to Question with Known Answer
6.7. Affirm
6.8. Deny/Reject
6.9. Other Tutorials

7. **Social Functions**
7.1. Talk Routines
7.2. Other Socials

8. **Procedural Functions**
8.1. Call
8.2. Availability Response
8.3. Contingent Query/Repetition
8.4. Contingent Query/Identification
8.5. Response to Contingent Query
8.6. Check
8.7. Affirm
8.8. Deny/Reject
8.9. Other Procedurals

9. **Speech or Acts for Self**
9.1. Commentary
9.2. Other Speech or Act for Self

10. **Mood Categories**

APPENDIX C
IACV Communication Functions

The definitions of Communication Functions for subsequences are derived from the work of Gordon Wells (1975, 1985). This appendix is an abstract of section 'The IACV Manual for Transcription and Analysis' (Van Balkom et al., 1990; section 5.3).

1. Subsequence or Conversational Theme

Within a topic of conversation smaller units of discourse may occur which realise subsidiary purposes within the overall purpose of the topic. So within a specific topic in which the child has to shut the door after entering the house, there may be an initiating (Procedural) subsequence to gain the child's attention, followed by a subsequence (Control) requesting her to shut the door; the child may interject an account of where he has been (Representational) followed by Mother's exclamation of surprise (Expressive) before she reverts to the original request and the child complies (Control).

2. Communication Functions of Subsequences

Within the hierarchically structured description of the instrumental nature of discourse we come to the level of Communication Functions, which describes the function that a particular verbal or nonverbal act performs.

The term 'Speech Act' has been used by philosophers such as Austin (1962) when talking about the purposes individual utterances serve. They have been concerned only with utterances in isolation, whereas, as we have seen, utterances do not usually occur in isolation, but more likely as moves within larger plans and in combination with nonverbal acts. Verbal and nonverbal acts, nevertheless, have a unique status as they are the smallest units of communicative interaction - the building blocks from which the edifice of conversation is constructed.

In describing the communicative functions of verbal and nonverbal acts, we are concerned with the purpose an act is intended to realise in the context in which it occurs, and more specifically with the purpose it conventionally realises within the language rather than with the more remote outcomes it may be intended to bring about.

So, for example, the utterance "Would you mind shutting the window?" spoken in an aggrieved tone by an elderly lady to her nephew is coded as a Command.

The fact that it is realised syntactically as an interrogation will be accounted for under the scoring category 'Mood' and the further fact that the tone in which it is uttered is intended by the elderly lady to make her nephew ashamed of his thoughtlessness in causing her to be in a draught, falls outside this coding scheme altogether. The complete list of the possible functions is given separately for each subsequence mode. An act occurring within a Control subsequence, for example, therefore must be coded by one of the functions listed under Control. If a function listed under a different subsequence mode seems more appropriate, is an indication that the subsequence as a whole has been incorrectly coded, or that a boundary has occurred, and that a new subsequence has been initiated.

Examples of the Change of Communication Functions

a.	where's your apple cake?	R
	<playing with dolls' house>	
c.	uh?	P
a.	where is the apple cake?	P
c.	//in there	R
c.	points to box on floor]	
a.	go and get it	C

---

a	: adult/caregiver
c	: child
//..]	: start (//) and end (]) of overlap
R	: Representational
P	: Procedural
C	: Control
<...>	: context

This is a Control sequence initiated by the mother. Her aim is to get the child to play with the apple cake. She begins with a Representational subsequence, asking where the cake is. The child however, does not understand her question and so switches to a Procedural subsequence in order to have the information repeated. The mother maintains this subsequence by repeating the question. The child is then able to respond to the original question and in so doing he reverts to the Representational subsequence. Then the mother commands the child to get the apple cake, thus realising a Control subsequence which is the dominant purpose of the sequence.

So, because of the changing communicative functions, all these acts are coded as initiatives in order to indicate that change of function.

3. Control Functions

The control of the present or future behaviour of one or more of the participants. This may concern a particular act or a general disposition to behave in a particular way, and so will include commands and requests for action as well as statements about what ought to be done, and supporting justifications. Within IACV the following Controls are defined:

>   General Request
>   Request for Justification
>   Request for Permission
>   Other Request
>   Wanting
>   Intend
>   Prohibition
>   Justification
>   Assent
>   Acknowledge
>   Reject/refuse
>   Other Controls

3.1. General Request

A request by the speaker/actor for action on the part of the addressee. Requests may take different syntactic forms (as will be coded in 'Mood') according to the relative status of the participants, the formality of the situation, etc.

>   *Examples of General Request*
>   - take the ball
>   - shall we read a book?
>   - will you stop playing now?
>   - take care of your watch
>   (warnings are coded as 'general requests' too)

3.2. Request for Justification

A request for a reason or justification for a Command, Promise, Treat, etc.

> *Examples of Request for Justification*
> - why not?
> - why are you crying?

3.3. Request for Permission

A request to be permitted to do something.

> *Examples of Request for Permission*
> - may I play with the dolls?
> - can I have some water please?

3.4. Other Request

A request different from the ones already mentioned.

3.5. Wanting

Acts that state a desire for a change to be brought about in the situation by another person usually for the speaker's benefit. This is one of the earliest functions to emerge, and later will tend to become differentiated into more specific functions. The act may, but need not, include "I want ...".

> *Examples of Wanting*
> - I want to drink something
> - another book!

3.6. Intend

Statement of the speaker's intention to do something, usually in a situation where the addressee may grant or withhold permission.

Examples of Intend
- I will be doctor now.
- we are going to play with the dolls.

3.7. Prohibition

A negative form of Command. Like a Command, it can be realised in various syntactic forms.

Examples of Prohibition
- don't touch!
- leave the phone alone!

3.8. Justification

Gives reason, explanation or justification to back up a Command, etc. or an act with some form of asserting function.

Example of Justification

1.a. don't do that
    <c is standing on book, on the floor>
2.a. it is not your book

---

act 1 is coded as 'Prohibition' and
act 2 is coded as 'Justification'

3.9. Assent

A positive response to an initiating Control Function only.

3.10. Acknowledgement

An act which simply signals that the preceding Control message has been received, or an action has been recognised.

Examples of Acknowledgement

1.a. //here a cookie
2.a. gives cookie to c.]
3.c. thank you mum

act 3 is coded as 'Acknowledgement'

1.a. close the door Linda
2.c. closes the door

nonverbal act 2 is an 'Acknowledgement'

a: adult/caregiver
c: child
//...]: start (//) and end (]) of overlap
<...>: context

3.11. Reject/refuse

A rejection of Commands, Promises, (e.g., on the grounds of their inappropriateness and irrelevance) or a negative response to an initiating Control Function only. Frequently a Refusal is linked with a Justification, and the Refusal may itself be ellipsed.

Example of Refuse

1.c. I want a cookie
2.a. you can't have a cookie now
3.a. we do not have any cookies here

Act 2 is a 'Refusal'
Act 3 is a 'Justification'

Example of Reject

<a. is playing with dolls>
1.c. //plays with dolls
2.a. let's go eating now]
3.c. I am playing with the dolls

Act 3 is coded as 'Reject'

---

a: adult/caregiver
c: child
//..]: start (//) and end (]) of overlap
<..>: context

3.12. Other Controls

Acts which cannot be described by the preceding Control Functions must be scored in this category.

4. Expressive Functions

The expression of feelings and attitudes as an affective response to a situation. This category is concerned with spontaneous reaction rather than with considered opinion. Expressive sequences are, therefore, normally rather brief and contain little structure. Within IACV the following Expressives are defined:

>   Exclamation
>   Accompaniment to Action
>   Query State or Attitude
>   Express State or Attitude
>   Encourage
>   Approval
>   Disapproval
>   Responses
>   Other Expressives

4.1. Exclamation

Outbursts of feeling, screaming etc. Usually one-word acts or phrases.

> *Examples of Exclamation*
> - OH that is big!
> - BAH!

4.2. Accompaniment to Action

Act that accompanies behaviour in an expressive way, or makes an expressive comment on that behaviour, or on the situation in which it occurs.

> *Example of Accompaniment to Action*
>
> <a. completes a puzzle>
> a. //DONE
> c. looks at a.]
> ---
> a: adult/caregiver
> c: child
> //..]: start (//) and end (]) of overlap
> <..>: context
> CAPITALS: stressed word(-part)
> ---

4.3. Query State or Attitude

Requesting the addressee to express his feelings or attitude. As with previous category, emphasis is on affective tone.

> Examples of Query State or Attitude
> - do you like that play?
> - did you hurt yourself?

4.4. Express State or Attitude

Statement about the speaker's feelings or attitudes. Such statements may include facts but are predominantly affective in tone. They differ from Exclamations in that they are much more likely to receive a response.

> Examples of Express State or Attitude
> - what a nice girl you are
> - you have made a mess again
> - isn't that a beautiful picture
> - I don't like tomato-soup

4.5. Encourage

Acts that encourage the addressee in his behaviour.

> Examples of Encourage
> - try again
> - come on
> - you know how to do it

4.6. Approval

Positive reinforcement of an action or an act, for example of an agreement to carry out a command or a response to a question in a Tutorial subsequence.

> *Examples of Approval*
> - okay
> - that is right
> - well done boy

4.7. Disapproval

Negative reinforcement of action or act.

> *Examples of Disapproval*
>
> - I'm very cross with you
> - that was a stupid thing to do

4.8. Responses

Responses to acts.

> *Example of Responses*
>
> 1.a. are you glad?
> 2.c. yes
> ---
> Act 2 is coded as 'Expressive response'
> ---
> a: *adult/caregiver*
> c: *child*

4.9. Other Expressives

Acts that cannot be described by the preceding Functions must be scored in this category.

5. Representational Functions

The exchange of information. Discursive discussion, including considered evaluation, of any aspect of experience is also covered by this category. Whereas with Control sequences there is the intention that the verbal and nonverbal acts lead to some eventual action, Representational speech does not have action as the intended outcome, although naturally all information could have implications for action. The expression of affective attitude also enters into most exchanges of information, but unless this is the dominant purpose of the conversation, it is the informational aspect that takes precendence and so the sequence is coded as Representational. Within IACV the following Representationals are defined:

> Ostension
> Statement
> Content Question
> Response to Content Question
> Yes/no Question
> Affirm

Deny/Reject
Other Question
Other Responses
Justification
Other Representationals

5.1. Ostension

Act designed to indicate or point out an object. Includes early naming acts.

Example of Ostension

<playing with doll's house>
1.a. //points to object at floor]
2.c. a car]

Act 2 is an 'Ostension'

---

a: adult/caregiver
c: child
//...]: start (//) and end (]) of overlap
<...>: context

5.2. Statement

Presentation of a piece of information (but not including Responses to Questions, Justifications). Statements are generally concerned with facts and include descriptions of internal states or attitudes when these are predominantly factual.

Examples of Statement
- there is a cow in the garden
- look there is a car
- I like playing with you

Onomatopoeses simultaneously performed with associated nonverbal acts are coded as statements.

Example of Statement/Onomatopoesis

1.c. //jumps with cat
 <cuddly toy>
2.c. miauw]
 <onomatopoesis>

Act 2 is coded as 'Statement'

'Rhetorical Questions' and 'Tag-Questions' are also coded as statements.

Example of Statement/Rhetorical Question/Tag-Question

1.c. sits on floor
2.a. you are going to sit on the floor?
 or
3.a. it is nice to sit on the floor, isn't it?

Act 2 is coded as a 'Statement/Rhetorical Question'
Act 3 is coded as a 'Statement/Tag Question'

---

a: *adult/caregiver*
c: *child*
//...]: *start (//) and end (]) of overlap*

5.3. Content Question

Requesting the addressee to provide a specific piece of information, as mostly is indicated by the 'wh'-word.

Examples of Content Question
- who was playing with that car?
- where did you get that from?

Questions to name something and 'Why'-Questions are not coded as 'Content Questions' but as 'Other Questions' (see section 5.8.).

5.4. Response to Content Question

Provision of specific piece of information in response to Content Question. Responses are frequently elliptical, but even complete utterances that are elicited by Content Questions are coded as Responses.

> *Example of Response to Content Question*
>
> <playing with the animal-farm>
> 1.a. where did you find that horse?
> 2.c. in the farm
> ---
> Act 1 is coded as a 'Content Question'
> Act 2 is coded as a 'Response to Content Question'
> ---
> a: *adult/caregiver*
> c: *child*
> <...>: *context*

5.5. Yes/No Question

Requesting the addressee to agree or disagree with a statement.

> *Examples of Yes/No Question*
> - is that car red?
> - that is an elephant?
> - have you got a dress for that doll?

5.6. Affirm

Positive response to Yes/No Question or confirmation of truth of statement or response.
    Positive reinforcements ('that is good'), compliments etc. are also coded as 'Affirm'.
The nonverbal act 'looking at' is coded as 'Affirm'.

Example of Affirm

<playing with animal-farm>
1.c. //puts animal into box
2.a. looks at activities c.]

Act 1 is a 'Statement'
Act 2 is an 'Affirm'

5.7. Deny/Reject

A negative response to Yes/No Question or a rejection of the truth or relevance of a preceding act or part of an act.

Example of Deny/Reject

<playing with animal farm>
1.a. that is a cow
2.c. //no that is a horse
3.c. nodding her head]

Act 1 is coded as a 'Statement'
Act 2 is a 'Deny/Reject'
Act 3 is a 'Deny/Reject'

---

a: adult/caregiver
c: child
//...]: start (//) and end (]) of overlap
<...>: context

5.8. Other Questions

Other questions. These can be "Why Questions" or questions to name something in cases where the answer is not known.

5.9. Other Responses

Responses to "Other Questions".

5.10. Justification

This function is much wider in scope in the Representational Mode since it includes all forms of supporting information to back up statements. It is also used in response to 'Why Questions'.

> *Example of Justification*
>
> <playing cards>
> 1.c. what is that?
> 2.a. that is a bird
> 3.a. an owl
>
> ---
>
> Act 3 is a 'Justification'
>
> ---
>
> *a: adult/caregiver*
> *c: child*
> *<...>: context*
>
> ---

5.11. Other Representationals

Acts that cannot be described by the preceding Representational Functions must be scored in this category.

6. Tutorial Functions

Interaction where one of the participants has a deliberately didactic purpose. Within IACV the following Tutorials are defined:

> Demand for Required Form
> Supply of Required Form
> Question with Frame
> Filler to Question with Frame

Question with Known Answer
Response to Question with Known Answer
Affirm
Deny/Reject
Other Tutorials

6.1. Demand for Required Form

The purpose is to teach or provide opportunity for practice of a particular form of act.

Examples of Demand for Required Form
- say how are you?
- say thank you

6.2. Supply of Required Form

Response to Demand for Required Form, whether correct or not. Incorrect responses are likely to be followed by Reject or Disapproval and/or a further Demand.

Example of Supply of Required Form

<playing cards>
1.a. say, cow
2.c. "cob"

Act 1 is a 'Demand for Required Form'
Act 2 is a 'Supply of Required Form'

---

a: *adult/caregiver*
c: *child*
<...>: *context*

6.3. Question with Frame

An act left incomplete with an intonation contour which indicates that the addressee is to supply the missing part.

6.4. Filler to Question with Frame

Supplying the missing part of a Question with Frame, whether the response is correct or not.

> *Example of Filler to Question with Frame*
>
> <playing with dolls' house>
> 1.a. mamma is in the-
>   <referring to doll>
> 2.c. //kitchen
> 3.c. looks into dolls' house]
>
> ---
>
> Act 1 is a 'Question with Frame'
> Acts 2 and 3 are 'Fillers to Question with Frame'
>
> ---
>
> a: *adult/caregiver*
> c: *child*
> <...>: *context*
> //...]: *start (//) and end (]) of overlap*
> -: *indication for unfinished part of act*

6.5. Question with Known Answer

A question posed by speaker who knows the answer but is either checking the addressee's ability to answer correctly or helping him to see an argument. The question may be either Pos/Neg or Wh-. The response will usually be followed by some form of evaluation.

6.6. Response to Question with Known Answer

(see preceding function).

> *Example of Question with Known Answer and Response*
>
> <playing with animal-farm>
> 1.a. how //do you call this one?
> 2.a. points at animal on floor]
>   <cow>

3.c. a cow
4.a. good

Acts 1 and 2 are 'Questions with Known Answer'
Act 3 is a 'Response to Question with Known Answer'
Act 4 is an 'Affirm'

6.7. *Affirm*

Positive response or confirmation of accuracy of a response.

6.8. *Deny/Reject*

A negative response to a question or a refusal to an answer or a rejection of an elicited response on grounds of inaccuracy or inappropriateness. Corrections are also coded as Deny/Reject.

Example of Deny/Reject

<playing with medical kit>
1.a. //what is this?
2.a. points at medical kit]
3.c. medpits
4.a. a medical kit

Act 4 is a 'Deny/Reject' (intended as correction)

---

a: *adult/caregiver*
c: *child*
<....>: *context*
//...]: *start (//) and end (]) of overlap*

6.9. *Other Tutorials*

Acts that cannot be described by the preceding functions must be scored in this category.

7. Social Functions

Conversation is mainly concerned with maintaining social relationships. In addition to greetings and ritualistic formulae, social sequences may be concerned with the weather and other conventionally agreed subjects. They also include such games as "peek-a-boo", the purpose of which is simply to enjoy social interaction. Within IACV the following Socials are defined:

> Talk Routines
> Other Socials

7.1. Talk Routines

Many families have verbal games that have evolved from earlier childish acts or pre-verbal interaction. Examples are games like "peek-a-boo", "Pat-a-cake" or repeated imitations by both participants. The criterion for assignment to this category is that the acts should have little or no referential significance and should follow a relatively fixed pattern. This category also includes singing and recitations in which several participants join together either simultaneously or alternately.

7.2. Other Socials

Acts that cannot be described by "Talk routines" must be scored in this category.

8. Procedural Functions

Procedural acts are concerned with the channel of communication rather than with its content. Such subsequences can occur to initiate or end a sequence, or at any time within a sequence to rectify a breakdown in communication due to mishearing or misunderstanding.
Within IACV the following Procedurals are defined:

> Call
> Availability Response
> Contingent Query/Repetition
> Contingent Query/Identification

Response to Contingent Query
Check
Affirm
Deny/Reject
Other Procedurals

8.1. Call

An act designed to gain attention before the speaker/actor embarks on the main part of his communication.

8.2. Availability Response

Response to call; it indicates that the speaker/actor is paying attention and ready to receive communication.

Example of Call and Availability Response

<a. wants c. to stop playing>
1.a. LINDA
2.a. HEE
3.c. yes I stop

Act 1-2 are 'Calls'
Act 3 is a 'Availability Response'

---

a: adult/caregiver
c: child
<...>: context
CAPITALS: stressed word(-part)

8.3. Contingent Query/Repetition

A request that the previous act should be repeated or reformulated because it has not been heard or understood.

Examples of Contingent Query/Repetition
- what did you say?
- again!
- I don't understand

8.4. Contingent Query/Identification

Identifying act in response to a request.

Examples of Contingent Query/Identification
- which one do you mean?
- is this the car you mentioned?

8.5. Response to Contingent Query

Response to "Repetition" and "Identification"

8.6. Check

A repetition or reformulation of the whole or part of the previous act with interrogative intonation, the purpose of which is to check that the previous act has been correctly understood.

Example of Check

<playing with dolls' house>
1.c. takes chair
2.c. //isse chaa
3.a. looks at chair]
   <in the child's hands>
4.a. is that a chair?

Act 4 is a 'Check'

---

a: *adult/caregiver*
c: *child*
<...>: *context*
//...]: *start (//) and end (]) of overlap*

230

8.7. Affirm

Confirms the correctness of an Identification or Check.

8.8. Deny/Reject

Deny of the correctness of an Identification, Check or Reject.

8.9. Other Procedurals

Acts that cannot be described by the preceding functions must be scored in this category.

9. Speech or Acts for Self

Since Ego-centric Speech is not primarily intended to have a communicative purpose it would be inappropriate to describe it in terms of the hierarchical structure of conversation. Ego-centric acts fulfil important functions for the speaker. However, they are dissimilar from the purposes conversational sequences serve in interpersonal interaction. Within IACV the following Speech or Acts for Self are defined:

>Commentary
>Other Speech or Act for Self

9.1. Commentary

Acts which either comment on ongoing activities or plan future action. Commentary may often act as a form of control or monitoring of non-verbal behaviour in the execution of larger complex tasks.

Example of Speech for Self/Commentary

<c. is playing with garage>
1.c. //moves elevator in garage to the third floor
2.c. one, two, three]
  <whispers>

Act 2 is Speech for Self/Commentary

---

c: child
<...>: context
//...]: start (//) and end (]) of overlap

9.2. Other Speech or Act for Self

Acts that cannot be described by "Commentary", must be scored in this category.
Watching activities of the addressee must also be scored in this category.

10. Mood Categories

Although Mood is a system of grammatical structure, it is convenient to include it in the part of the analysis that deals with Communicative Functions, as the options it provides allow a further subclassification of Function. The subfunction Command, for example, can be realised by verbal acts in the different moods described below:

"I want you to shut the door"	(Declarative)
"Would you mind shutting the door?"	(Interrogative)
"Shut the door"	(Imperative)
"The door!"	(Moodless)

Which of the syntactic variants of Command is chosen depends largely on the speaker's assessment of the relative status of the participants and the formality of the situation. Similar choices from the Mood system operate for many functions and in general it would be true to say that there is no simple one-to-one correspondence between the function of an utterance and its syntactic form. This can give rise to ambiguity, for example where a Command is realised by an interrogative sentence (Mother to child: "Can you make a little less noise?"). However in normal conversation such occasions rarely arise, as the context provides the adressee with enough clues to decide which is the intended function: one of the basic assumptions underlying conversation is that the speaker's utterances will be relevant, and where the adressee does not see the immediate relevance, he must look for premises that would allow a relevant interpretation.

APPENDIX D

Interjudge Reliability Measures for the IACV Pragmatic Categories

This Appendix contains an overview of the reliability measures of the IACV Pragmatic Categories and are based on Cohen's (1960) Kappa.

Cohen's Kappa is based on calculations from two independent coders, who did double codings for the IACV Pragmatic Categories using 12% of the total IACV data (18 transcripts; 12 transcripts from different SLI children and sessions; 6 transcripts from different NLA children and sessions).

Cohens's Kappa's are given for the various Pragmatic Categories and for the Complete IACV Pragmatic Codings.

(See also section 5.3.2.)

The transcripts which were coded twice by two independent coders* are:

SLI Dyads (DN = Dyad Number)
(1) (DN 12)   Remco Session 1
(2) (DN 3)   Rianne Session 1
(3) (DN 8)   Bjorn Session 4
(4) (DN 2)   Saskia Session 4
(5) (DN 11)   Christian Session 6
(6) (DN 4)   Davy Session 6

---
* The original coders

(7) (DN 5)   Lisette Session 6
(8) (DN 6)   Geronimo Session 7
(9) (DN 10)   Johny Session 7
(10) (DN 7)   Marcel Session 7
(11) (DN 9)   Sebastian Session 7
(12) (DN 1)   Linda Session 9

NLA Dyads (DN = Dyad Number)
(13) (DN 15)   Tim Session 1
(14) (DN 14)   Bram Session 3
(15) (DN 13)   Rachel Session 4
(16) (DN 17)   Rosanna Session 6
(17) (DN 16)   Michel Session 7
(18) (DN 18)   Chantalle Session 9

APPENDIX D
Interjudge Reliability Measures for the IACV Pragmatic Categories
(based on Cohen's (1960) Kappa)

Table D-1   Reliability Measures IACV

IACV Category	Cohen's Kappa*	Confidence Limits (95%)
Mutual Utterance Relation (Initiations, Responses, Comments)	0.72	0.67 - 0.75
Simultaneous Relation (overlaps, parallel talk)	0.89	0.86 - 0.92
Correctness of Interaction (content and use correct and incorrect)	0.95	0.92 - 0.96
Direction and Relation of Acts (partner, self or other)	0.89	0.86 - 0.91
Feedback (backchannels, clarification requests)	0.78	0.75 - 0.81
Imitations (partner acts, self acts)	0.98	0.97 - 0.99
Elaboration Previous Act (suppletion, correction)	0.96	0.95 - 0.98
Topic Organisation (introduction, continuation, finishing)	0.98	0.97 - 0.99
Communication Function (Controls, Tutorials, Representationals)	0.65	0.60 - 0.72
Total IACV Pragmatics	0.78	0.74 - 0.83

* Based on double codings of 12% of the total data (18 transcripts of different dyads; a total of 2824 acts of caregivers and children)

APPENDIX E

Mean and Standard Deviations of Communication Function Categories of the Caregivers

This Appendix contains 5 Tables with the Group Mean and Standard Deviations for the various subcategories of Controls, Tutorials, Expressives, Representationals and Procedurals of the 12 caregivers in the SLI dyads and the 6 caregivers in the NLA dyads.

The Mean scores are given first for each subcategory and the Standard Deviations are put below the Mean, between brackets (<...>).

The Tables E-1 to E-5 belong to chapter 7 (section 7.5.2.).

APPENDIX E
Mean and Standard Deviations Subcategories of IACV Communication Functions of Caregivers in the SLI and NLA Dyads

TABLE E-1: Controls Caregiver

Subcategories	SLI Caregivers	NLA Caregivers
Requests	8.41 <SD 3.68>	6.59 <SD 1.47>
Wanting/Intend	20.48 <SD 9.16>	13.72 <SD 4.68>
Prohibition	2.24 <SD 1.22>	0.78 <SD 0.22>
Justification	4.89 <SD 2.27>	4.69 <SD 1.19>
Assent/Acknowledge	11.72 <SD 5.58>	10.63 <SD 2.51>
Refuse/Reject	2.89 <SD 1.59>	1.69 <SD 0.23>
Other Controls	1.59 <SD 0.58>	0.94 <SD 0.48>

TABLE E-2: Tutorials Caregiver

Subcategories	SLI Caregivers	NLA Caregivers
Demands for Required Form	3.43 <SD 2.42>	0.64 <SD 0.22>
Supply of Required Form	2.43 <SD 1.68>	0.19 <SD 0.09>
Question with Frame	4.43 <SD 2.48>	0.44 <SD 0.21>
Filler to Question with Frame	1.33 <SD 1.19>	0.17 <SD 0.08>
Question Known Answer	5.65 <SD 4.69>	2.26 <SD 1.05>
Response to Question Known Answer	1.57 <SD 0.51>	0.37 <SD 0.15>
Affirm	2.69 <SD 1.60>	0.63 <SD 0.13>
Deny/Reject	2.13 <SD 0.60>	0.26 <SD 0.12>
Other Tutorials	2.17 <SD 0.36>	0.31 <SD 0.13>

TABLE E-3: Expressives Caregivers

Subcategories	SLI Caregivers	NLA Caregivers
Exclamation	2.17 <SD 0.96>	0.57 <SD 0.45>
Accompaniment to Action	1.79 <SD 0.96>	0.44 <SD 0.29>
Query State/Attitude	1.44 <SD 0.36>	0.31 <SD 0.12>
Express State/Attitude	1.70 <SD 0.67>	0.94 <SD 0.60>
Encourage	3.00 <SD 0.94>	0.17 <SD 0.07>
Approval	1.92 <SD 0.37>	0.63 <SD 0.33>
Disapproval	1.37 <SD 0.28>	0.09 <SD 0.12>
Responses	1.30 <SD 0.22>	0.07 <SD 0.15>
Other Expressives	1.17 <SD 0.19>	0.07 <SD 0.17>

TABLE E-4: Representationals Caregivers

Subcategories	SLI Caregivers	NLA Caregivers
Ostension	3.16 <SD 1.47>	0.74 <SD 0.32>
Statements	21.27 <SD 10.12>	15.17 <SD 4.42>
Content Question	6.92 <SD 1.22>	5.52 <SD 0.22>
Response Content Question	2.96 <SD 1.16>	3.11 <SD 0.42>
Yes/No Question	10.08 <SD 3.95>	11.19 <SD 2.51>
Affirm	7.21 <SD 2.88>	6.74 <SD 1.53>
Deny/Reject	3.04 <SD 1.42>	1.72 <SD 0.65>
Other Questions	1.69 <SD 0.64>	0.44 <SD 0.21>
Other Responses	1.83 <SD 0.53>	1.28 <SD 1.18>
Justification	4.96 <SD 2.57>	3.67 <SD 0.87>
Other Representationals	1.43 <SD 0.43>	1.06 <SD 1.24>

TABLE E-5: Procedurals Caregiver

Subcategories	SLI Caregivers	NLA Caregivers
Call	2.88 <SD 1.54>	0.83 <SD 0.45>
Availability Response	1.64 <SD 0.49>	0.50 <SD 0.26>
Contingent Queries	4.11 <SD 1.24>	0.76 <SD 070>
Response to Contingent Queries	2.06 <SD 0.58>	0.56 <SD 0.49>
Check	3.29 <SD 1.77>	2.85 <SD 1.63>
Deny/Reject	1.91 <SD 0.70>	0.24 <SD 0.13>

APPENDIX F

IACV Tables of Statistics

This Appendix presents the Tables of Statistics (Mean (M) and Standard Deviations (SD)) of the codings on IACV categories for the 12 SLI and 6 NLA dyads for the nine sessions referred to in the chapters 6 and 7.

The tables contain the session mean and standard deviations of the 12 SLI dyads and 6 NLA dyads. The (individual) subject mean and standard deviations (for each dyad over the complete period of 18 months) and the overall mean and standard deviations (over all sessions and all dyads) are not given in this book. That information is available at the Institute for Rehabilitation Research, Hoensbroek.

KEY:

SLI : *Specific Language Impaired Group*
NLA : *Normal Language Acquiring Group*
S-1 ... S-9: *Session 1 .... Session 9*
T : *Total*
M : *Mean*
SD : *Standard Deviations*

APPENDIX F
Frequency Tables belonging to the Chapter 6 and 7

TABLE F1a MEAN (M) AND STANDARD DEVIATIONS (SD) OF THE TOTAL ACTS USED

	\multicolumn{9}{c}{TOTAL ACTS USED}		
	S-1	S-2	S-3	S-4	S-5	S-6	S-7	S-8	S-9	M	SD
NLA DYADS											
T	1262	1380	1360	1443	1322	1349	1392	1361	1302	—	—
M	210.33	213.33	226.67	240.50	220.33	224.83	231.83	226.83	217.00	223.52	9.42
SD	45.77	23.46	21.88	22.89	25.69	19.87	24.41	23.24	38.82	27.34	8.80
SLI DYADS											
T	2956	2924	2965	3117	3210	3251	3242	4380	3238	—	—
M	246.29	243.58	247.04	259.75	267.50	270.88	270.17	264.96	269.83	260.00	35.39
SD	51.86	66.23	66.57	63.27	65.67	65.83	55.14	54.78	63.40	49.62	16.76

T : TOTAL ACTS   M : MEAN   D : STANDARD DEVIATIONS

TABLE F1b MEAN (M) AND STANDARD DEVIATIONS (SD) OF VERBAL AND NONVERBAL ACTS USED

TOTAL NONVERBAL ACTS CHILD

	S-1	S-2	S-3	S-4	S-5	S-6	S-7	S-8	S-9	M	SD
NLA CHILD											
M	59.15	61.17	64.83	64.83	66.50	67.00	62.17	65.67	56.67	63.11	3.5
SD	8.16	14.85	15.54	3.92	16.55	11.80	6.11	15.59	18.66	12.35	5.1
SLI CHILD											
M	64.00	67.42	70.17	76.08	86.83	82.42	92.75	81.58	79.50	77.86	9.3
SD	18.37	31.16	33.73	31.22	31.91	12.93	25.33	21.79	15.36	25.67	6.6

TOTAL VERBAL ACTS CHILD

	S-1	S-2	S-3	S-4	S-5	S-6	S-7	S-8	S-9	M	SD
SLI CHILD											
M	59.33	53.50	55.83	62.50	47.00	55.00	55.17	53.83	53.83	55.11	4.2
SD	10.79	15.64	22.85	18.74	19.73	17.37	10.54	14.77	13.54	16.00	4.1
NLA CHILD											
M	53.25	58.17	64.08	60.75	65.33	60.92	69.08	65.25	62.00	62.09	4.6
SD	17.87	13.37	17.33	15.98	9.25	7.67	10.61	12.89	9.04	12.67	3.7

M : MEAN   SD : STANDARD DEVIATIONS

TABLE F2 MEAN (M) AND STANDARD DEVIATIONS (SD) OF IACV-CATEGORIES BELONGING TO 'TURN-TAKING' HYPOTHESES

	\multicolumn{9}{c}{TURN-TAKING HYPOTHESES}		
	S-1	S-2	S-3	S-4	S-5	S-6	S-7	S-8	S-9	M	SD
BACKCHANNELS ADULT (NLA GROUP)											
M	5.67	5.17	9.67	6.00	8.00	9.00	6.33	7.83	8.17	7.31	4.18
SD	4.32	5.71	4.23	2.28	6.96	5.73	3.50	3.54	3.43	1.97	1.27
BACKCHANNELS ADULT (SLI-GROUP)											
M	6.54	7.88	7.58	9.96	10.67	6.71	8.38	10.08	8.75	8.50	3.02
SD	2.50	3.32	3.81	3.46	2.96	3.25	3.28	4.02	3.49	2.04	.73
BACKCHANNELS NLA-CHILD											
M	.67	2.50	1.83	.17	.67	1.17	.50	1.00	1.17	1.07	1.51
SD	1.03	2.88	1.72	.41	.82	1.17	.55	1.67	1.94	.39	.61
BACKCHANNELS SLI-CHILD											
M	2.82	3.54	3.73	3.75	3.71	3.95	5.45	4.83	5.00	3.90	1.99
SD	1.75	2.39	2.91	3.04	3.24	2.71	2.90	1.95	2.52	1.86	.90
ELLIPSIS NLA-CHILD											
M	7.83	5.76	5.25	8.39	9.16	4.70	12.22	6.33	7.98	7.51	2.33
SD	6.79	7.39	5.88	6.92	3.60	3.43	3.92	9.50	4.37	5.76	2.08
ELLIPSIS SLI-CHILD											
M	12.75	6.67	8.67	13.92	6.33	9.33	16.00	4.58	8.33	9.62	3.82
SD	7.83	8.67	8.17	7.33	5.83	5.83	5.17	7.33	5.67	6.87	1.26

TURN-TAKING HYPOTHESES

	S-1	S-2	S-3	S-4	S-5	S-6	S-7	S-8	S-9	M	SD
MEAN LENGTH OF TURN ADULT (NLA-GROUP)											
M	1.84	2.10	2.21	2.10	2.27	1.88	2.22	2.14	2.19	2.10	.15
SD	.26	.43	.58	.35	.67	.21	.63	.30	.65	.45	.18
MEAN LENGTH OF TURN ADULT (SLI-GROUP)											
M	2.44	2.21	2.43	2.34	2.51	2.12	2.05	2.47	2.50	2.34	0.52
SD	0.61	0.43	0.45	1.05	0.86	0.45	0.40	0.43	0.67	0.34	0.24
ELABORATION THEME ADULT (NLA-GROUP)											
M	38.33	46.17	48.67	54.17	47.17	42.50	52.50	51.17	54.33	48.33	21.36
SD	10.19	14.37	20.28	25.93	21.10	26.04	35.38	14.16	42.18	8.66	11.04
ELABORATION THEME ADULT (SLI-GROUP)											
M	79.50	44.67	39.00	60.08	42.92	42.67	63.00	38.58	39.92	50.04	14.26
SD	28.14	13.44	22.29	18.25	13.64	27.43	27.47	11.07	16.94	19.85	6.69
REINITIATIONS ADULT (NLA-GROUP)											
M	1.67	2.83	2.00	3.83	2.33	.50	1.67	4.83	2.67	2.48	2.21
SD	2.42	3.19	1.26	5.19	2.58	.84	1.21	4.88	2.88	2.00	1.50
REINITIATIONS ADULT (SLI-GROUP)											
M	2.25	7.42	3.17	1.33	6.50	2.25	2.33	7.92	1.92	3.90	2.60
SD	2.01	4.91	2.59	2.10	3.92	2.01	2.27	6.91	2.02	3.19	1.73

	S-1	S-2	S-3	S-4	S-5	S-6	S-7	S-8	S-9	M	SD
TURN-TAKING HYPOTHESES											
SELF-RELATED IMITATIONS BY ADULT (NLA-GROUP)											
M	3.50	4.67	3.17	2.67	4.67	3.17	4.33	4.50	5.50	4.02	2.50
SD	2.95	2.42	1.72	1.51	3.88	3.13	3.01	2.51	2.81	1.32	.24
SELF-RELATED IMITATIONS BY ADULT (SLI-GROUP)											
M	12.58	7.75	6.42	6.83	5.67	4.50	6.50	3.17	3.58	6.33	4.53
SD	9.37	4.18	4.06	4.90	2.93	2.20	4.01	2.37	2.35	1.85	2.38
PARTNER-RELATED IMITATIONS BY NLA-CHILD											
M	1.67	3.17	2.83	2.00	1.83	1.50	3.00	2.50	1.83	2.26	1.68
SD	1.21	2.64	1.17	1.26	3.54	1.22	2.45	2.43	1.17	1.08	.78
PARTNER-RELATED IMITATIONS BY SLI-CHILD											
M	3.00	4.42	3.50	2.75	3.17	2.92	2.08	2.33	1.58	2.86	2.28
SD	1.60	3.78	3.61	2.56	3.56	2.02	1.78	3.65	1.78	1.70	0.97
NONVERBAL INITIATIONS NLA-CHILD											
M	25.33	21.17	26.67	23.00	24.50	24.00	22.67	22.67	26.83	24.09	11.37
SD	10.58	9.28	15.12	14.90	10.19	15.68	12.68	11.38	19.31	6.44	3.23
NONVERBAL INITIATIONS OF SLI-CHILD											
M	30.68	23.33	33.57	36.73	30.92	42.69	42.96	32.23	32.14	33.94	12.95
SD	14.15	9.30	12.67	14.02	10.27	12.62	12.15	11.61	19.92	7.82	4.03

TABLE F3   MEAN (M) AND STANDARD DEVIATIONS (SD) OF IACV-CATEGORIES BELONGING TO 'CHILD ADJUSTED REGISTER (CAR)'

	\'CHILD ADJUSTED REGISTER (CAR)\' HYPOTHESES										
	S-1	S-2	S-3	S-4	S-5	S-6	S-7	S-8	S-9	M	SD
MLU ADULT (NLA-GROUP)											
M	5.13	5.62	5.34	5.74	5.60	4.99	5.13	5.49	5.65	5.41	.27
SD	1.25	.98	.68	.98	.97	1.13	1.54	.91	.90	1.04	.24
MLU ADULT (SLI-GROUP)											
M	4.49	4.56	4.63	4.81	4.88	4.93	4.86	5.18	5.33	4.85	.28
SD	.76	1.32	.80	1.00	.91	1.12	.60	.76	.93	.91	.21
MLU NLA-CHILD											
M	3.62	3.82	3.80	4.19	4.48	3.44	3.86	3.81	4.62	3.96	.39
SD	.72	.47	.37	.54	.83	.69	.92	.77	1.14	.72	.24
MLUSLI-CHILD											
M	1.72	2.50	2.74	2.64	2.96	3.07	2.81	3.29	3.26	2.78	.48
SD	.47	.84	.89	.89	.86	1.19	.57	.86	1.06	.85	.22
REQUEST FOR INFORMATION BY ADULT (NLA-GROUP)											
M	15.50	16.00	19.17	16.50	15.33	20.67	20.33	9.50	21.33	17.15	7.66
SD	6.09	3.35	10.36	3.73	10.07	10.17	12.40	5.92	11.34	4.78	2.64
REQUEST FOR INFORMATION BY ADULT (SLI-GROUP)											
M	20.00	15.00	20.46	18.38	14.29	19.46	18.46	12.33	12.08	16.72	6.38
SD	9.63	6.79	6.63	5.33	5.00	7.96	6.95	2.86	6.33	3.38	2.05

TABLE F4 MEAN (M) AND STANDARD DEVIATIONS (SD) OF IACV-CATEGORIES BELONGING TO 'TOPIC AND THEME MANAGEMENT'

	S-1	S-2	S-3	S-4	S-5	S-6	S-7	S-8	S-9	M	SD
INTRODUCTION NEW THEME BY ADULT (NLA GROUP)											
M	3.67	4.00	3.00	3.17	2.33	2.17	1.83	2.00	3.50	2.85	2.36
SD	2.42	3.58	5.06	1.17	2.07	1.83	1.72	1.79	2.51	1.01	1.07
INTRODUCTION NEW THEME BY ADULT (SLI GROUP)											
M	3.08	1.42	2.33	3.67	2.50	1.83	3.00	3.00	1.33	2.46	.81
SD	2.50	1.08	2.64	2.10	1.62	1.19	1.81	1.65	1.37	1.78	.55
INTRODUCTION NEW THEME BY NLA-CHILD											
M	4.33	3.83	5.33	5.67	2.83	2.17	3.00	3.67	3.83	3.85	2.82
SD	3.44	2.14	4.03	4.23	2.14	1.94	1.79	3.08	2.32	1.03	.65
INTRODUCTION NEW THEME BY SLI-CHILD											
M	4.25	2.17	2.92	4.58	3.67	3.75	6.00	3.42	2.42	3.69	1.18
SD	3.47	1.85	2.54	3.20	2.06	2.63	3.05	1.98	2.43	2.58	.57
MEAN LENGTH OF SUBSEQUENCE ADULT (NLA-GROUP)											
M	2.98	2.55	2.55	2.77	2.55	2.76	3.98	2.99	2.47	2.73	.46
SD	1.70	1.30	1.43	1.11	1.41	2.91	1.30	.45	1.22	1.43	.65
MEAN LENGTH OF SUBSEQUENCE ADULT (SLI-GROUP)											
M	1.96	2.59	2.34	2.09	2.74	2.42	2.09	2.82	3.26	2.48	0.58
SD	0.28	0.65	0.63	0.26	0.79	0.83	0.20	0.68	1.30	0.50	0.31

	\multicolumn{9}{c	}{Topic and Theme Management}		
	S-1	S-2	S-3	S-4	S-5	S-6	S-7	S-8	S-9	M	SD
MEAN LENGTH OF SUBSEQUENCE NLA-CHILD											
M	2.76	2.73	3.36	3.28	3.02	4.18	2.81	2.67	2.81	3.07	.48
SD	1.23	.76	2.92	2.46	1.32	3.69	1.25	.77	1.28	1.74	1.03
MEAN LENGTH OF SUBSEQUENCE SLI-CHILD											
M	2.58	3.17	2.70	2.56	3.04	2.44	2.38	3.39	3.97	2.91	0.73
SD	0.39	0.74	0.74	0.37	0.65	0.60	0.38	0.96	1.97	0.61	0.44

TABLE F5 MEAN (M) AND STANDARD DEVIATIONS (SD) OF IACV-CATEGORIES BELONGING TO 'COMMUNICATION FUNCTIONS (COMFUN)

	\multicolumn{9}{c}{COMFUN NLA CAREGIVERS}		
	S-1	S-2	S-3	S-4	S-5	S-6	S-7	S-8	S-9	M	SD
Controls											
M	29.50	35.50	35.83	48.50	39.17	34.83	46.67	55.67	40.83	40.72	8.17
SD	10.56	9.97	16.20	17.76	17.57	16.29	20.41	13.72	32.43	17.21	6.64
Expressives											
M	3.17	5.33	1.83	1.50	2.67	2.33	5.83	5.17	2.00	3.31	1.68
SD	3.82	3.08	3.13	2.26	4.23	4.76	7.73	6.97	3.35	4.37	1.85
Representationals											
M	48.67	47.17	56.50	49.67	47.50	60.67	60.00	30.83	54.67	50.63	9.07
SD	24.87	13.67	17.34	12.55	19.61	16.72	20.22	9.13	18.57	16.96	4.67
Tutorials											
M	5.17	6.50	6.83	8.67	4.50	3.33	.83	6.83	1.50	4.91	2.62
SD	5.08	5.86	9.22	11.98	5.92	2.50	2.04	12.02	1.22	6.21	4.09
Procedurals											
M	6.00	6.17	5.33	4.67	10.17	2.83	1.83	11.00	6.50	6.06	3.00
SD	8.58	8.57	5.35	6.31	10.34	5.23	2.23	6.96	7.50	6.79	2.37

	\multicolumn{9}{c	}{COMFUN NLA CHILDREN}		
	S-1	S-2	S-3	S-4	S-5	S-6	S-7	S-8	S-9	M	SD
Requests for Information											
M	12.67	12.17	9.33	9.50	8.00	11.00	11.33	9.33	6.83	10.02	4.76
SD	6.50	9.45	10.05	7.45	3.35	7.32	8.76	8.04	6.68	6.23	.96
Controls											
M	52.33	46.33	54.17	65.00	56.83	41.50	58.83	72.50	48.00	55.06	9.61
SD	17.13	21.92	24.55	9.01	22.09	23.90	11.34	25.38	31.93	20.81	7.19
Expressives											
M	2.17	5.17	2.00	.33	.67	1.67	3.00	3.17	3.17	2.37	1.47
SD	2.86	5.19	2.45	.52	.82	2.73	2.28	2.48	4.02	2.59	1.44
Representationals											
M	56.83	52.00	53.50	53.67	41.83	73.67	52.50	29.33	50.50	51.54	11.81
SD	28.69	25.32	16.15	23.22	15.09	26.64	15.68	7.26	16.38	19.38	6.96
Tutorials											
M	2.50	4.50	6.33	2.83	2.50	1.50	.50	4.00	.67	2.81	1.89
SD	2.81	3.89	8.16	3.06	2.66	1.38	1.22	7.01	.82	3.45	2.56
Procedurals											
M	3.50	3.17	3.00	3.33	4.67	2.33	1.83	7.50	5.00	3.81	1.70
SD	5.82	3.37	2.37	3.61	3.56	3.61	2.14	5.75	4.82	3.90	1.32

	S-1	S-2	S-3	S-4	S-5	S-6	S-7	S-8	S-9	M	SD
COMFUN SLI CAREGIVERS											
Controls											
M	56.75	46.92	42.67	44.71	43.71	43.38	41.13	38.83	40.00	44.23	12.53
SD	16.52	21.79	20.65	17.77	20.65	19.54	17.06	14.13	17.54	13.42	6.61
Expressives											
M	2.91	2.64	7.06	5.29	3.71	5.06	4.50	3.10	2.33	3.91	2.40
SD	1.38	1.72	7.15	5.75	2.54	2.03	2.06	1.68	1.15	2.17	2.09
Representationals											
M	50.79	48.58	56.88	55.33	58.08	60.58	54.54	56.58	65.42	56.31	14.93
SD	22.92	17.37	15.55	12.55	21.64	24.25	14.89	22.96	40.29	16.33	8.13
Tutorials											
M	11.96	14.17	9.23	5.92	9.29	10.42	9.59	12.41	22.57	10.89	8.06
SD	10.49	11.26	6.44	5.49	9.30	10.57	10.96	15.10	31.42	7.79	7.34
Procedurals											
M	7.08	9.73	6.23	8.83	10.29	5.32	6.00	6.88	6.17	7.42	3.85
SD	4.85	6.75	6.68	5.12	4.26	3.90	3.61	2.96	5.15	3.22	1.47

	COMFUN SLI CHILDREN									
	S-1	S-2	S-3	S-4	S-5	S-6	S-7	S-8	S-9	M	SD
Requests for information										
M	8.63	9.00	10.17	11.67	10.13	8.88	8.79	8.25	8.17	9.30	4.31
SD	9.90	7.80	8.80	9.28	8.60	4.65	3.17	5.60	8.46	6.11	2.35
Controls										
M	60.71	56.83	63.08	66.29	57.33	68.83	67.92	54.38	53.92	61.03	16.20
SD	23.90	23.25	23.61	16.48	11.80	12.67	15.33	13.73	24.64	10.15	6.56
Expressives										
M	3.23	3.27	6.38	6.38	5.50	7.77	6.38	4.00	3.33	5.34	2.90
SD	1.81	2.94	4.22	4.07	4.85	5.28	4.04	2.52	1.51	2.60	1.73
Representationals										
M	41.75	50.92	57.33	55.33	64.83	64.79	53.71	55.67	64.92	56.58	16.71
SD	15.54	23.86	25.91	19.90	30.02	20.35	5.48	15.56	27.85	14.06	8.17
Tutorials										
M	6.13	7.92	5.86	5.15	7.09	7.45	7.50	8.56	14.57	6.92	4.86
SD	6.79	6.15	5.91	3.41	6.48	7.06	9.78	9.62	18.46	5.04	4.59
Procedurals										
M	6.64	8.09	5.35	6.50	7.71	5.50	5.54	7.00	6.75	6.45	3.40
SD	6.39	6.35	5.76	5.48	4.62	3.61	4.00	5.83	4.43	3.64	1.91

TABLE F6 MEAN (M) AND STANDARD DEVIATIONS (SD) OF IACV-CATEGORIES BELONGING TO 'COMMUNICATION BREAKDOWNS AND REPAIRS'

	\multicolumn{11}{c}{Communication Breakdown and Repairs}
	S-1	S-2	S-3	S-4	S-5	S-6	S-7	S-8	S-9	M	SD
UNFINISHED VERBAL ACTS NLA-CHILD											
M	2.00	2.50	1.17	2.83	2.00	1.17	1.17	1.50	1.67	1.78	1.35
SD	2.00	1.87	1.17	1.17	.89	.75	.75	1.38	2.25	.58	.52
UNFINISHED VERBAL ACTS SLI-CHILD											
M	1.67	2.55	3.75	2.11	3.09	2.56	3.57	4.22	3.00	2.80	1.53
SD	1.03	2.25	2.55	1.27	2.26	1.51	1.81	1.86	1.51	1.03	0.71
PARALLEL TALK IN NLA-DYADS											
M	1.00	2.33	1.83	4.00	3.33	2.33	1.67	2.00	5.00	2.61	1.87
SD	1.10	1.75	1.47	2.76	2.66	2.34	1.51	1.79	3.74	1.47	1.03
PARALLEL TALK IN SLI-DYADS											
M	2.00	3.08	3.50	2.90	4.00	2.80	2.00	3.55	4.10	2.98	1.70
SD	1.32	1.68	2.72	3.11	3.37	1.99	0.89	1.63	2.33	1.28	0.89
CLARIFICATION REQUESTS ADULT (NLA-GROUP)											
M	4.00	3.33	3.17	2.17	3.67	2.17	2.33	1.50	4.17	2.94	1.87
SD	2.19	1.63	.98	1.94	2.16	1.60	2.34	1.76	3.60	1.14	.70
CLARIFICATION REQUESTS ADULT (SLI-GROUP)											
M	6.33	3.46	5.83	6.58	5.13	7.42	6.08	3.27	2.75	5.37	2.67
SD	2.48	1.59	2.16	2.38	3.85	6.00	3.23	1.78	1.83	1.97	1.47

	Communication Breakdown and Repairs										
	S-1	S-2	S-3	S-4	S-5	S-6	S-7	S-8	S-9	M	SD

CORRECTIONS OF PREVIOUS ACT BY ADULT (NLA-GROUP)
M	.50	.50	.33	.83	.00	1.00	.33	.17	1.00	.52	.98
SD	1.22	.84	.52	1.33	.00	2.00	.52	.41	1.26	.25	.46

CORRECTIONS OF PREVIOUS ACT BY ADULT (SLI-GROUP)
M	4.50	0.75	0.75	1.42	0.50	1.67	3.00	0.58	0.92	1.56	2.01
SD	3.03	1.76	1.06	1.73	0.80	1.56	3.46	0.67	1.08	0.87	0.93

CONSECUTIVE INITIATIONS IN NLA-DYADS
M	29.12	19.24	31.22	42.80	17.37	35.46	41.94	13.01	49.53	31.08	12.62
SD	24.10	10.52	25.55	28.05	5.65	35.02	18.93	10.22	34.60	21.41	10.77

CONSECUTIVE INITIATIONS IN SLI-DYADS
M	82.21	52.50	77.75	78.46	54.79	87.13	85.17	50.92	46.92	68.43	24.88
SD	22.12	20.07	22.30	18.66	24.09	30.04	11.63	20.47	45.22	16.73	6.29

FAULTY RESPONSES (PARTNER-RELATED) OF ADULT (NLA-GROUP)
M	1.00	.67	.83	1.33	2.67	1.17	1.17	1.83	1.50	1.35	1.18
SD	1.26	1.03	.98	.52	1.63	1.17	1.17	1.47	2.81	.89	.54

FAULTY RESPONSES (PARTNER-RELATED) OF ADULT (SLI-GROUP)
M	2.00	2.14	2.00	2.40	1.86	1.25	1.17	1.75	9.00	1.64	0.64
SD	1.29	1.21	0.00	1.52	1.21	0.50	0.41	1.84	0.00	0.92	0.88

	Communication Breakdown and Repairs										
	S-1	S-2	S-3	S-4	S-5	S-6	S-7	S-8	S-9	M	SD
FAULTY RESPONSES (PARTNER-RELATED) OF NLA-CHILD
M	2.00	3.83	2.00	1.17	4.83	1.83	.67	3.50	1.67	2.39	2.29
SD	3.52	2.99	2.37	1.17	2.64	1.83	1.21	2.51	3.14	1.41	.88
FAULTY RESPONSES (PARTNER-RELATED) OF SLI-CHILD
M	2.41	2.72	1.17	2.21	2.17	1.20	1.89	1.81	1.33	2.05	0.96
SD	1.55	1.66	0.41	1.25	1.09	0.45	1.05	1.00	0.58	0.66	0.65
FAULTY INITIATIONS ADULT (NLA-GROUP)
M	1.83	3.17	1.50	5.00	3.00	1.67	2.00	3.00	2.00	2.57	1.11
SD	1.83	1.33	1.38	2.28	2.83	1.03	1.90	2.10	2.37	1.89	.57
FAULTY INITIATIONS ADULT (SLI-GROUP)
M	5.63	7.58	3.75	6.17	7.17	3.58	7.50	7.63	3.58	5.84	1.78
SD	2.31	4.24	2.88	4.01	4.14	2.60	4.71	3.85	2.11	3.43	.96
FAULTY INITIATIONS NLA-CHILD
M	4.00	5.67	5.83	5.17	7.50	3.67	5.33	9.67	4.50	5.70	1.87
SD	3.52	2.66	3.87	3.19	3.83	2.42	2.58	7.53	4.93	3.84	1.59
FAULTY INITIATIONS SLI-CHILD
M	7.63	7.25	3.83	8.67	9.71	4.92	8.71	8.46	3.67	6.98	2.27
SD	4.05	3.71	2.82	3.20	4.13	2.55	3.48	3.98	3.42	3.48	.55

APPENDIX G

Early Social Communicative Scales (ESCS)

The Early Social Communicative Scales (ESCS) are developed by Seibert and Hogan (1982).
The authors of the ESCS gave permission to translate the scales and the caregiver interview protocol into Dutch.

The Dutch ESCS was used during the first home-visit made by the speech therapist of the IACV-project team in order to get additional information about the communicative functioning in the earlier stages.

The ESCS version given in this Appendix is the Dutch adaptation, used in the IACV study.

VROEG SOCIAAL-COMMUNICATIEVE SCHAAL

J.M. Seibert en A.E. Hogan (1981)

Datum :
Kind :
Observator :

IKV / IACV vertaling met toestemming auteurs

	0	niveau 1	niveau 2	niveau 3 — 3,0 — 3,5 — niveau 4

1.1. RSI / Respons op Sociale Interactie

1. kan gekalmeerd worden
2. lacht op horen stem volwassene (2x)
3. kalmeert na optillen
4a. enthousiast bij nadering volwassene
4b. strekt armen uit naar uitgestrekte armen volwassene
5. gebruikt eenvoudige a) vocale, b) non-vocale responsen of c) kijkt volwassene aan als volwassene stopt met vocalisaties of speletje spelen (2x)
6. kijkt weg van vreemde(n) of afwachtend/waakzaam
7. kijkt volwassene aan bij horen naam
8. herhaalt aandachttrekkende activiteit
9. a) vocale of b) nonvocale verzoek voor herhaling van vocalisatie(s), gebaren, speletje volwassene (2x)
   c) neemt deel aan imitatieve handeling passend in een speletje met volwassene
10. geeft voorwerp terug in door volwassene geïnitieerd voorwerp-uitwisselingsspelletje
11. beantwoordt groet volwassene
12. gebruikt conventioneel gebaar ter verandering van speletje van/met volwassene
13. antwoordt met niet-geïnitieerd conventioneel gebaar op gecombineerd vocaal/non-vocaal verzoek van volwassene tot "gesprekje"
14. troost de volwassene als deze voorwendt te huilen
15. gebruikt niet-geïnitieerde conventionele gebaren als reactie op initiatief voor gesprek door volwassene
16. gebruikt eenwoordsuiting (niet-geïnitieerd) als reactie op groet/initiatief tot "gesprek" van de volwassene
17. doet mee aan simulatiespelletje met voorwerp, vocaal voorgesteld door volwassene
18. gebruikt tweewoordsuitingen als antwoord op uitnodiging tot "gesprek" door volwassene

1.2. ISI / Initiatieven tot Sociale Interactie

1. kijkt spontaan naar gezicht volwassene
2. kijkt naar persoon die langs loopt/beweegt
3. reikt naar volwassene die nabij is, maar zelf geen contact zoekt/maakt
4. vraagt aandacht van volwassene
   a) door vocalisaties en oogcontact
   b) door gebaren met geluiden en oogcontact
   c) door vocalisaties, gebaren of lachen naar een toekijkende volwassene
5. a) kijkt en vocaliseert naar niet-sprekende, toekijkende volwassene
   b) kijkt en gebaart naar niet-sprekende, toekijkende volwassene
   c) kijkt, lacht en vocaliseert of gebaart naar toekijkende volwassene
6. groet spontaan d.m.v. conventioneel gebaar
7. geeft via conventioneel gebaar te kennen aan volwassene dat speletje begint
8. begint met voorwerp-uitwisselingsspeletje met volwassene
9. gebruikt conventioneel gebaar samen met volwassene
10. plaagt/daagt uit volwassene met bepaalde handelingen
11. gebruikt eenwoord-uitingen als initiatief voor a) groeten
    b) sociaal speletje (samen spelen)
    c) samen spelen met voorwerp
    d) uitwisselingsspeletje (voorwerpen)
    e) plagen/uitdagen
12. roept volw. buiten de kamer bij naam
13. gebruikt tweewoordsuiting als initiatief voor: a) groeten, b) sociaal spel (samen spelen), c) samen spelen met voorwerp, d) uitwisselingsspeletje (voorwerpen), e) plagen/uitdagen
14. begint een "doe-alsof" speletje, waarin volw. een rol krijgt toebedeeld

1.3. BSI / Bestendigen van Sociale Interactie

1. huilt/verzet zich bij neerzetten/-leggen door volwassene
2. gebruikt eenvoudige a) vocale of b) nonvocale reacties of c) maakt oogcontact steeds als handelingen (incl. vocalisaties) van volwassene stoppen (3x)
3. speelt "kiekeboe" door doek van gezicht volwassene te trekken (2x)
4. huilt als volwassene dreigt weg te gaan
5. bij onderbreken van zowel vocalisaties, gebaren en spelletjes door volwassene volgen reacties als: a) vocale, b) non-vocale uitingenreeks die door kind is aangezet c) neemt drie beurten m.b.v. een geïnitieerde conventionele handeling tijdens een sociaal speletje met de volwassene
6. neemt drie beurten in een a) vocale, b) non-vocale uitingenreeks die door kind is aangezet
7. neemt drie beurten tijdens uitwisselingsspelletje
8. plaagt/daagt uit (3x)
9. gebruikt niet-geïnitieerd conventioneel gebaar om sociaal spel of samenspel met voorwerp te bestendigen
10. gebruikt niet-geïnitieerde conventionele gebaar samen-handelen te bestendigen (2x) binnen minstens drie beurten, waarin a) groeten of b) samen spel, c) samenspel met voorwerp, d) uitwisselingsspeletje van voorwerpen of e) plagen/uitdagen centraal staat
11. neemt drie beurten in "doe-alsof" dialoog, speletje met volw. waarbij een niet-aanwezig voorwerp of een vervanging daarvan centraal staat
12. gebruikt tweewoordcombinaties in speldialoog in tenminste drie beurten m.b.t.: a) plagen, b) sociaal spel (samenspelen), c) samenspel met voorwerp, d) uitwisselingsspelspelletje van voorwerpen, e) plagen/uitdagen

VROEG SOCIAAL-COMMUNICATIEVE SCHAAL

J.M. Seibert en A.E. Hogan (1981)

2. AANDACHT

Datum : _____
Kind : _____
Observator : _____

2.1. RA / Reacties op Aandacht

Niveau	Item
0	1. kijkt naar voorwerp dat volwassene met nadruk laat zien
niveau 1	2. kijkt naar voorwerp dat volw. terloops laat zien (3x)
	3a. volgt wijsrichting wijsvinger volw. (2x)
niveau 2	3b. volgt de punt van uitgestrekte arm van volw. (3/4x)
	4. volgt de a) kijkrichting en/of b) wijsrichting van volw. in zowel linker als rechter kant (gelijkelijk verdeeld) na extra aanduiding daartoe
niveau 3	5. volgt constant a)kijk- en/of b)wijsrichting naar rechts of links (3/4x)
3,0	6. volgt a)kijk- en/of b)wijsrichting volwassene tot 180° achter kind
	7. antwoordt met hoofdbeweging (ja/nee) op vragen volw. over voorwerpen, plaatjes of gebeurtenissen
	8. kijkt/geeft aan a) twee of meer lichaamsdelen, b) bekende voorwerpen op gejkte plaatsen, c) bekende personen nadat volw. ze heeft genoemd
	9. pakt/wijst aan twee of meer genoemde voorwerpen uit groep voorwerpen
3,5	10. correct antwoord op "wat is dat..?"-vragen (twee/meer voorwerpbenoemteken)
	11. gebruikt eenwoorduiting om antwoord te geven op vragen van volw. over voorwerpen, personen en gebeurtenissen
	12. gebruikt eenwoordsuitingen om helderheid te verschaffen over bewering van volwassene
niveau 4	13. a) pakt 10 of meer voorwerpen op benoeming van een presentatieblad of b) wijst ze aan in een boek
	14. geeft vijf of meer voorwerpen/plaatjes aan op verzoek
	15. vindt verborgen voorwerp op benoeming
	16. gebruikt twee woorden als antwoord op vragen over voorwerpen, personen of gebeurtenissen
	17. gebruikt twee woorden om helderheid te krijgen over bewering volw.

2.2. IA / Initiatieven voor Aandacht

Niveau	Item
niveau 1	1. kijkt spontaan tijdens spel met voorwerp naar volw. (2x)
niveau 2	2a. kijkt afwisselend van een door de hand voortbewogen voorwerp naar volw., die ermee speelt
	2b. en lacht naar volw.
	3a. kijkt afwisselend van mechanisch voortbewegen voorwerp naar volw., terwijl deze het voorwerp activeert
	3b. en lacht naar volw.
	4. kijkt en wijst voorwerpen en plaatjes zonder volw. aan te kijken a) kortbij b) veraf
niveau 3	5. laat spontaan zien / geeft voorwerp aan volw., alleen om het te laten zien
	6. wijst spontaan naar voorwerp/plaatje en kijkt volw. aan
	7. benoemt voorwerpen/plaatjes spontaan en kijkt naar volw.
	8. gebruikt een woord als "aandachttrekker" a) voor voorwerp/plaatje b) ter beschrijving van voorwerp plaatje/situatie
niveau 4	9. vraagt benaming voorwerp / plaatje persoon
	10. gebruikt twee-woorduitingen als a) aandachttrekker voor voorwerp / plaatje b) beschrijving voor situatie of plaatje c) vraagstelling naar situatie, gebeurtenis, voorwerp of plaatje

2.3. AB / Aandacht Bestendigen

Niveau	Item
niveau 1	1. kijkt achtereenvolgens naar 3 voorwerpen die volw. actief laat zien of aanwijst
	2. kijkt achtereenvolgens naar 3 voorwerpen die volw. passief laat z.en
	3. kijkt spontaan 3x naa volw. terwijl het kind ergens mee speelt
niveau 2	4. kijkt afwisselend naar volw. en naar een voorwerp waar iets mee gedaan wordt (met de hand of mechanisch)
	5. kijkt naar een serie voorwerpen of plaatjes die een volw. aanraakt en waar een volw. naar wijst
	6. bij drie opeenvolgende mondelinge pogingen: kijkt afwisselend naar twee voorwerpen en reikt naar een van de twee objecten
	6x. volgt 3x achter elkaar een punt waar de volw. met gestrekte arm naar wijst
niveau 3	7. volgt 3x achter elkaar de kijk- of wijsrichting van een volw.
	8. wijst 3x naar plaatje in een boek en kijkt naar volw. (hij mag ertoe aangezet worden om te wijzen)
	9. als de volw. zomaar met het kind plaatjes of voorwerpen bekijkt, wijst het k:nd ook min. 1x iets aan
	10. kijkt of wijst naar drie a) lichaamsdelen, b) voorwerpen op een vertrouwde plaats, c) bekende personen als volw. die achter elkaar opnoemt
	11. wijst correct naar minstens 3 plaatjes of voorwerpen die volw. benoemt
	12. gebruikt 3x één-woord-uitingen om iets te benoemen of plaatjes te beschrijven spontaan of in antwoord op een vraag, terwijl hij tenminste 1x oogcontact maakt met volw.
niveau 4	13. wijst correct naar tenminste 10 plaatjes of voorwerpen die een volwassene noemt
	14. benoemt tenminste 5 voorwerpen of plaatjes achter elkaar, spontaan of na aansporing
	15. vraagt naar de naam van 3 verschillende voorwerpen
	16. maakt tenminste één twee-woord-zin als commentaar op een voorwerp of gebeurtenis bij een verbale interactie van tenminse 3 beurten
	17. antwoordt op een serie vragen van volwassene en zegt spontaan iets over gebeurtenissen die voorbij zijn (3x)

IRV / IACV vertaling met toestemming auteurs

VROEG-SOCIAAL-COMMUNICATIEVE SCHAAL

J.M. Seibert en A.E. Hogan (1981)

Kind : _____
Observator : _____
Datum : _____

niveau 4	3,5	niveau 3	3,0	niveau 2	niveau 1	0
				3.1. RGR / Reacties op Gedrags Regulatie		
1. draait in richting stem volw. (3x) 2. stopt bezigheid na "nee" met aanraken (1 of 2x) 3. verzet zich tegen het wegnemen van een voorwerp door een volw. 4. wendt zich af of duwt een ongewenst voorwerp weg als dit in contact komt met zijn lichaam		5. stopt bezigheid na "nee" of ander verbaal commando (1 of 2x) 6. duwt/slaat de hand van een volw. weg als deze een voorwerp weg pakt 7. protesteert/verzet zich tegen wegnemen voorw. door volw. en zoekt oogcontact 8. volgt één van de aanwijzingen van niveau 3 op		9. begrijpt minstens 2 verschillende verbale aanwijzingen gecombineerd met een gebaar a. geef het aan mij f. blijf zitten b. leg het erin g. kom op, we gaan c. pak het af h. kom hier d. kus de baby i. ga zitten e. trek eens aan x. schudt "nee" na een opdracht 10. begrijpt tenminste 2 verbale aanwijzingen (situatie-gebonden) a. geef het aan mij i. zeg eens dag b. leg het erin j. kus de baby c. pak het af k. geef de baby eten d. kus de baby l. kam het haar van de baby e. trek eens aan m. laat de baby slapen g. ga zitten n. geef de baby te drinken h. kom hier 11. gebruikt een woord om speeltje te weigeren dat volw. vraagt of probeert te pakken en/of verzet zich tegen commando 12. volgt een reeks van drie tegengestelde aanwijzingen, waarbij handeling en voorw. veranderen bij elk commando, zoals bijv.: a. voedt baby e. laat baby slapen b. kus het hondje f. haasje slaapt c. geef baby drinken g. hondje zit d. laat hondje springen 13. gebruikt tweewoordszinnetjes om speeltje te weigeren dat volw. vraagt of probeert te pakken en/of verzet zich tegen commando				
				3.2. IBR / Initiatieven om uw gedrag te reguleren		
				5. vertoont tenminste 2 van de volgende gedragingen: A. het verkrijgen van voorwerpen a) reikt aanhoudend naar een voorwerp buiten bereik (2x) b) maakt oogcontact met volw. als het voorw. buiten bereik is (2x) x) reikt naar voorwerp en maakt afwisselend oogcontact met volw. (2x) c) reageert als volw. een voorwerp laat zien B. Bewegend voorwerp d) maakt oogcontact met volw. als bewegend voorw. stopt met bewegen (2x) e) reikt naar de hand van volw. als bewegend voorw. stopt 6. vertoont één gedrag van niveau 3 7. vertoont tenminste 2 van de volgende gedragingen: A. a) "vraagt" naar een voorw. dat buiten bereik is gezet b) "vraagt" naar een niet verplaatsbaar voorw., buiten bereik B. c) geeft voorw. terug of gebruikt vaststaand gebaar met oogcontact als volw. stopt met spelen d) geeft voorw. aan volw. terug als hij wil dat het opgewonden wordt e) geeft voorw. aan volw. voordat getoond is hoe het werkt 8. gebruikt één woord in 2 verschillende situaties of twee verschillende woorden in dezelfde situaties a) volw. geeft een voorwerp g) weigert aangeboden object b) voorw. wordt buiten bereik gezet h) geeft voorw. aan volw. om ervan af te zijn c) voorw. staat buiten bereik i) heeft hulp nodig bij het rangschikken van voorw. of het verwijderen v.e. obstakel d) volw. stopt met spelletje j) wil toestemming (alsjeblieft) e) voorw. stopt met spelen k) wil dat een vervelende situatie ophoudt f) het gebruik van het voorw. is nog niet gedemonstreerd l) andere hulpvragende activiteit 9. vraagt naar 2 verschillende voorwerpen die hij niet kan zien 10. gebruikt een tweewoordsuiting om te vragen naar: a) een voorwerp b) een schouwspel c) iedere andere hulpvragende activiteit	C. Overige f) duwt een ongewenst voorw. weg voor het hem aanraakt, geen oogcontact met volw. g) geeft voorw. aan volw. als hij boos of angstig is, of pijn heeft h) kijkt naar volw. of raakt hem aan om activiteit te stimuleren i) laat merken dat iets moet ophouden C. f) weigert aangeboden voorw. terwijl hij kijkt naar volw. g) geeft voorw. aan volw. om er van af te zijn h) roept hulp in om voorw. te rangschikken of obstakels te verwijderen i) vraagt toestemming j) andere hulpvragende activiteiten	

IRV / IACV vertaling met toestemming auteurs

SAMENVATTING

VROEG SOCIAAL COMMUNICATIEVE VAARDIGHEDEN
behorend bij OUDERVRAGENLIJST IACV

	Reacties op Sociale Interactie (RSI)	Initiatief voor Sociale Interactie (ISI)	Bestendigen van Sociale Interactie (BSI)	Reacties op Aandacht (RA)	Initiatief voor Aandacht (IA)	Bestendigen van Aandacht (BA)	Reacties op Gedragsregulatie (RGR)	Initiatief voor Gedragsregulatie (IRG)
NIVEAU 0 Responsief 0-2 mnd.	1	geen items	geen items	1	geen items	geen items	geen items	geen items
NIVEAU 1 Eenvoudig, vrijwillig niet-gedifferentieerd 2-7 mnd.	2 3 4 5	1 2 3	1 2	2	1	1 2 3	1 2 3 4	1 2 3 4
NIVEAU 2 Complex, Gedifferentieerd 7-13 mnd.	6 7 8 9	4 5	3 4 5 6	3 4	2 3 4	4 5 6	5 f g h i 6 7 8	5- a b c d e f g h i 6
NIVEAU 3 Regulatie door terugkoppeling 13-21 mnd.	10, 11, 12 13, 14 ___ 3,0 15 16 ___ 3,5	6, 7, 8, 9, 10 11- a b c d e	7, 8, 9 10- a b c d e	5, 6, 7 8, 9, 10, 11, 12	5 6 7, 8- a b 9 10- a b c	7, 8, 9 10, 11, 12 13, 14, 15	7- a b c d e f g h i 8- a b c d e f g h i j k 9- a b c d e f g h i 10- a b c d e f g h i 11	7- a b c d e f g h i 8- a b c d e f g h i j k 9 10- a b c
NIVEAU 4 Anticipator Regulatie 21+ mnd.	17 18	12 13- a b c d e 14	11 12 a b c d e	13, 14 15, 16 17	9 10- a b c	13, 14, 15 16	12 13	9 10- a b c

Vaardigheidsniveau _____ Mediaal niveau: _____ Optimaal niveau: _____

Vaard.Ontw. niveau _____

Algemeen gemiddeld niveau: _____

Algemene ontwikkelingsleeftijd (benadering): * _____ (gemiddelde)

IACV / l.j.m. van balkom
juni 1985

© Seibert & Hogan, 1982